# Pickling and Fermentation for Preppers

**Nourishment for Challenging Times**

**Jackson Davis**

© Copyright 2024 - All rights reserved.

The content contained within this book may not be reproduced, duplicated or transmitted without direct written permission from the author or the publisher.

Under no circumstances will any blame or legal responsibility be held against the publisher, or author, for any damages, reparation, or monetary loss due to the information contained within this book, either directly or indirectly.

### Legal Notice:

This book is copyright protected. It is only for personal use. You cannot amend, distribute, sell, use, quote or paraphrase any part, or the content within this book, without the consent of the author or publisher.

### Disclaimer Notice:

Please note the information contained within this document is for educational and entertainment purposes only. All effort has been executed to present accurate, up to date, reliable, complete information. No warranties of any kind are declared or implied. Readers acknowledge that the author is not engaging in the rendering of legal, financial, medical or professional advice. The content within this book has been derived from various sources. Please consult a licensed professional before attempting any techniques outlined in this book.

By reading this document, the reader agrees that under no circumstances is the author responsible for any losses, direct or indirect, that are incurred as a result of the use of information contained within this document, including, but not limited to, errors, omissions, or inaccuracies.

# Table of Contents

**INTRODUCTION** .................................................. 6

**CHAPTER I: Introduction to Pickling and Fermentation** .... 8

   Definition of pickling and fermentation ......................... 8

   History and significance of pickling and fermentation . 11

   Benefits of pickling and fermentation for preppers ..... 14

   Overview of what the book will cover ......................... 16

**CHAPTER II: Understanding the Science Behind Pickling and Fermentation** .......................................................... 20

   Explanation of microbial fermentation ........................ 20

   Role of beneficial bacteria and yeasts ........................ 23

   Chemical processes involved in pickling and fermentation .......................................................... 25

   Factors affecting fermentation (temperature, salt concentration, pH) ..................................................... 28

**CHAPTER III: Essential Equipment and Ingredients** ........ 31

   Tools needed for pickling and fermentation ................ 31

   Types of containers for fermentation ........................... 34

   Essential ingredients (vegetables, salt, water, spices) .. 37

   Safety precautions and sanitation practices ................ 40

**CHAPTER IV: Pickling Techniques** ................................ 43

   Basic pickling methods (vinegar-based) ...................... 43

   Quick pickling vs. traditional pickling .......................... 46

Recipes for pickled vegetables (cucumbers, carrots, beets) ............... 49

Tips for achieving desired texture and flavor ............... 52

## CHAPTER V: Fermentation Basics ............... 55

Introduction to lacto-fermentation ............... 55

Creating a fermentation brine ............... 57

Fermentation vessels and their uses ............... 60

Commonly fermented vegetables (sauerkraut, kimchi) 63

## CHAPTER VI: Advanced Fermentation ............... 67

Exploring wild fermentation ............... 67

Using starter cultures and fermentation cultures ............... 70

Fermented beverages (kombucha, kefir) ............... 73

Troubleshooting common fermentation problems ............... 75

## CHAPTER VII: Health Benefits of Pickling and Fermentation ............... 79

Nutritional advantages of pickled and fermented foods ............... 79

Gut health and probiotics ............... 81

Immune system support ............... 84

Weight management benefits ............... 87

## CHAPTER VIII: Pickling and Fermentation for Long-Term Storage ............... 91

Strategies for preserving food during emergencies ............... 91

Proper storage techniques for pickled and fermented foods ............... 94

Shelf life and safety considerations .............................. 96

Incorporating pickled and fermented foods into emergency rations .......................................................... 99

## CHAPTER IX: Incorporating Pickling and Fermentation into Prepper Lifestyles ................................................... 103

Integrating pickling and fermentation into meal planning .................................................................................. 103

Creating resilient food systems .................................... 105

Community-building through food preservation ........ 108

Recipes and meal ideas for preppers ........................... 111

## CHAPTER X: Beyond the Basics: Creative Applications . 115

Pickling and fermenting unconventional ingredients . 115

Fusion recipes incorporating pickled and fermented foods .............................................................................. 118

Crafting homemade condiments and sauces .............. 120

Innovations in pickling and fermentation technology ..123

## CONCLUSION .............................................................. 127

# INTRODUCTION

In an age marked by uncertainty and upheaval, the value of self-sufficiency and resilience in food preparation has never been more apparent. "Pickling and Fermentation for Preppers: Nourishment for Challenging Times" is a comprehensive guide designed to empower individuals with the knowledge and skills necessary to harness the transformative power of pickling and fermentation, particularly in the context of preparedness for unforeseen challenges.

As the world grapples with disruptions to food supply chains, natural disasters, and other crises, preserving and storing food effectively becomes paramount. Pickling and fermentation, age-old methods of food preservation, offer not only a means of extending the shelf life of perishable foods but also a way to enhance their nutritional value and flavor profile. This book serves as a roadmap for those seeking to embrace these ancient culinary techniques as part of their preparedness strategy.

The journey begins with exploring the science behind pickling and fermentation, delving into the fascinating microbial processes that transform raw ingredients into tangy, probiotic-rich delights. Readers will gain a more profound comprehension of the role of beneficial bacteria and yeasts, as well as the factors that influence successful fermentation, from salt concentration to pH levels.

With this foundational knowledge, readers will be guided through the essential equipment and ingredients needed to embark on their pickling and fermentation journey. From traditional Mason jars to specialized fermentation crocks, the options for fermenting vessels are diverse, ensuring accessibility for preppers of all levels of expertise. Safety precautions and sanitation practices are emphasized, underscoring the importance of maintaining hygienic conditions throughout fermentation.

Within the pages of this book, readers will discover a wealth of practical techniques and recipes for pickling and fermenting an array of vegetables, fruits, and even beverages. Whether quick pickling crisp cucumbers or fermenting spicy kimchi, each method is accompanied by clear instructions and helpful tips to ensure success.

Beyond the practical aspects of food preservation, "Pickling and Fermentation for Preppers" explores the myriad health benefits of pickled and fermented foods. From supporting gut health to bolstering the immune system, these age-old preservation methods offer many advantages that are especially relevant in times of crisis.

As readers delve deeper into pickling and fermentation, they will discover creative ways to incorporate these preserved delights into their everyday meals, from zesty relishes to tangy sauces. The book concludes with a look toward the future of food preservation, exploring emerging trends and technologies that promise to further revolutionize how we prepare and store food.

In essence, "Pickling and Fermentation for Preppers" is more than just a cookbook—it is a roadmap to self-sufficiency, resilience, and nourishment in the face of adversity. Whether you are an experienced fermenter or a novice prepper, this book offers something for everyone, empowering readers to take control of their food supply and embrace the timeless art of pickling and fermentation.

# CHAPTER I

# Introduction to Pickling and Fermentation

## Definition of pickling and fermentation

Pickling and fermentation are two ancient methods of food preservation that have been used for thousands of years across various cultures worldwide. Despite their long history, these techniques continue to be popular in contemporary kitchens, not only for their preservative qualities but also for their ability to enhance flavors and nutritional value. This section explores the definitions, processes, and distinctions between pickling and fermentation, shedding light on their unique characteristics and their roles in food science and culinary traditions.

Fundamentally, pickling is the process of preserving food by submerging it in an acidic solution, usually vinegar, or by fermenting the brine, which produces acid, primarily lactic acid. The acidic environment created is inhospitable to many bacteria, slowing spoilage and preserving food. This technique can be applied to various foods, including vegetables, fruits, meats, and eggs. Pickling preserves the food and imparts a distinct sour flavor, which is highly valued in many culinary traditions. The critical component in pickling, the acidic solution, can be customized with various spices, herbs, and flavorings to create a diverse array of pickled products, each with its unique taste and culinary application.

Fermentation, conversely, is a metabolic process that produces chemical changes in organic substrates through the action of enzymes. In food, fermentation refers to converting carbohydrates to alcohols, carbon dioxide, and organic acids using microorganisms—yeasts or bacteria—under anaerobic conditions. Humans have harnessed fermentation to produce various foods and beverages, including bread, cheese, yogurt, beer, and wine. Unlike pickling, which primarily uses an externally applied acidic solution for preservation, fermentation relies on natural or added microorganisms to produce the acids necessary for preservation. This process not only extends the shelf life of the food but also enhances its nutritional value, as fermentation can increase the availability of vitamins and minerals, improve digestibility, and generate probiotics, which are beneficial for gut health.

While pickling and fermentation are used for preserving food, they do so through different mechanisms, resulting in products with distinct flavors, textures, and nutritional profiles. Pickling, relying on an acidic solution, often leads to a crisp texture and a tangy flavor, making pickled foods excellent as condiments, side dishes, or snacks. Fermentation, due to the activity of microorganisms, can produce a range of flavors and textures, from the creamy tang of yogurt to the bubbly lightness of fermented beverages, and the complex flavors of fermented vegetables like kimchi and sauerkraut. Furthermore, the fermentation process can create bioactive compounds and promote the development of beneficial bacteria, contributing to the food's health benefits.

The distinction between pickled and fermented foods is not always clear-cut, as some pickled foods undergo fermentation to develop their acidic environment, while some fermented foods may be pickled for additional flavor and preservation. This overlap showcases the versatility of these preservation methods and their ability to complement each other in culinary applications.

In terms of cultural significance, both pickling and fermentation hold important places in food traditions around the world. Pickled foods are central to cuisines from Asia to the Americas, serving as crucial components in meals for their ability to add depth and contrast to dishes. Similarly, fermented foods are celebrated for their complex flavors and health benefits, with traditional fermented products like kefir, kombucha, and miso being integrated into modern diets for their taste and nutritional advantages.

In conclusion, pickling and fermentation are time-honored food preservation techniques offering more than just extended shelf life. They are deeply ingrained in culinary traditions, providing a means to enhance flavors, improve nutritional content, and maintain the integrity of foods. While they utilize different processes—pickling through the addition of acid and fermentation through the metabolic activities of microorganisms—they share the goal of preserving food in a way that celebrates taste, tradition, and health. As modern culinary practices continue to evolve, the principles of pickling and fermentation remain relevant, underscoring the enduring importance of these methods in connecting us to our cultural heritage and to the fundamental joy of eating well-preserved, flavorful food.

## History and significance of pickling and fermentation

The history and significance of pickling and fermentation are as rich and diverse as the cultures that have developed and nurtured these practices over millennia. These methods of food preservation, while serving the practical purpose of prolonging the shelf life of perishable items, have also played a crucial role in the cultural, nutritional, and culinary development of societies around the globe. This section explores the historical roots of pickling and fermentation, their evolution over time, and their enduring significance in contemporary food practices and traditions.

The practices of pickling and fermenting food dates back to ancient times, when prehistoric societies realized how important it was to stockpile food for times of scarcity. The earliest archaeological evidence of fermentation dates back to around 7000 BCE in Jiahu, China, where residues of a fermented beverage made from rice, honey, and fruit were found. Similarly, pickling can be traced back to ancient Mesopotamia around 2400 BCE. These methods were not only practical solutions to the problem of food preservation but also contributed to the dietary variety, enhancing flavors and nutritional value.

The importance of these practices extends beyond mere preservation. Fermentation and pickling have deep cultural and religious significance in many societies. For instance, wine has played a central role in Christian rituals, while kimchi has been a staple in Korean cuisine for centuries, signifying the importance of family and tradition. These practices were also essential for survival, enabling sailors and explorers to take long voyages. Fermented foods like sauerkraut were crucial for preventing scurvy among sailors during long sea voyages by providing a vital source of Vitamin C.

Over the centuries, the art and science of pickling and fermentation have evolved, influenced by geographical, cultural, and technological factors. The spread of these techniques can be traced along trade routes, with knowledge and microbial cultures being exchanged alongside goods and spices. This led to the diversification of fermented and pickled foods, with each region developing its own unique flavors and methods. For example, the production of cheese and yogurt became prevalent in Europe, kimchi and soy sauce in Asia, and pickled vegetables and meats in Eastern Europe.

The Industrial Revolution and the advent of modern food preservation methods in the 19th and 20th centuries led to a decline in traditional pickling and fermentation practices. However, there has been a resurgence of interest in these methods in recent decades, driven by a growing awareness of their health benefits, including improved gut health and increased nutrient availability. This revival has been marked by a renaissance in artisanal and craft food production, where traditional techniques are valued for the depth and complexity of flavor they impart to food.

Today, pickling and fermentation are celebrated for their ability to bridge past and present, offering a tangible link to our cultural heritage while providing health benefits and flavor enhancements to modern diets. Fermented foods like kefir, kombucha, and sourdough bread are enjoyed for their probiotic qualities, while pickled vegetables are appreciated for their tangy crunch and flavor. These practices are also being explored for their potential in sustainable food systems, as methods to reduce waste by preserving seasonal produce and utilizing parts of foods that would otherwise be discarded.

The significance of pickling and fermentation extends beyond the kitchen and dining table. These practices represent a form of cultural expression, embodying the history, traditions, and environmental adaptations of the people who have developed and refined them. They are a testament to human ingenuity in harnessing natural processes for food preservation, flavor development, and nutritional enhancement.

Moreover, the communal aspect of pickling and fermentation, where knowledge and cultures are shared within communities, underscores the role of these practices in social cohesion and cultural continuity. Workshops, festivals, and community events centered around pickling and fermentation serve not only to educate but also to bring people together, fostering a sense of connection and shared heritage.

In conclusion, the history and significance of pickling and fermentation are multidimensional, encompassing practical, nutritional, cultural, and social aspects. These age-old customs have changed throughout time, adapting to new contexts and technologies, yet they remain deeply rooted in the human experience. As we continue exploring the flavors, health benefits, and environmental potentials of fermented and pickled foods, we also celebrate the rich cultural tapestries these practices weave. They remind us of our collective ingenuity and the enduring importance of food as a medium of cultural expression and connection.

## Benefits of pickling and fermentation for preppers

For individuals who identify as preppers—those preparing for possible future emergencies or disruptions in societal norms—the methods of pickling and fermentation hold significant value. These age-old food preservation techniques are not just culinary arts but essential skills for ensuring food security and nutritional diversity in times of uncertainty. This section delves into the myriad benefits that pickling and fermentation offer to preppers, from extending the shelf life of perishables to enhancing nutritional content, improving food safety, and fostering self-sufficiency.

At the heart of prepping is the emphasis on self-reliance and readiness for various scenarios, including natural disasters, economic downturns, or other crises that might disrupt the food supply chain. In such contexts, the ability to preserve food becomes paramount. Pickling and fermentation stand out as particularly effective methods for several reasons. Firstly, they significantly extend the shelf life of foods. Vegetables, fruits, meats, and even dairy products can be preserved for months or years, mitigating the risk of food shortages. This durability is crucial for preppers, who aim to maintain a stable food supply over long periods.

Beyond mere preservation, these techniques also enhance the nutritional value of food. Fermentation, in particular, is known to increase the bioavailability of nutrients, which makes it easier for the body to absorb vitamins as well as minerals. Foods like sauerkraut and kimchi are rich in vitamins C and K, as well as in probiotics, which are beneficial for digestive health. This aspect of fermentation is especially valuable in situations where fresh produce might be scarce, ensuring that preppers can maintain a balanced diet with the nutrients necessary for good health.

Safety is another significant benefit. The acidic environment created during pickling and the competitive exclusion of harmful bacteria by beneficial microbes during fermentation reduce the risk of foodborne illnesses. This natural form of food safety is significant in emergency scenarios, where access to medical care might be limited. By relying on these methods, preppers can ensure that their food supply is long-lasting and safe to consume.

Moreover, pickling and fermentation support the goal of self-sufficiency, a core principle for preppers. These techniques can be applied with basic tools and ingredients at home, reducing dependence on commercial food products and supply chains. This autonomy is crucial in emergencies, where access to grocery stores or markets might be restricted. Additionally, mastering these methods allows preppers to adapt to available resources, preserving whatever produce they can grow or forage, thus enhancing their resilience in various scenarios.

Pickling and fermentation are also noteworthy for their sustainability. These environmentally friendly methods require no electricity or advanced technology, making them ideal for off-grid living or situations where energy resources are scarce. Moreover, by allowing for the preservation of seasonal produce, these practices help reduce food waste, aligning with sustainable living principles that are often important to the prepper community.

Culturally, the practice of pickling and fermentation connects individuals to traditional foodways and community knowledge. Sharing recipes and techniques within prepping communities or with future generations fosters a sense of connection and continuity, which can be psychologically beneficial in times of crisis. This cultural aspect reinforces community bonds and ensures the preservation of valuable skills that might otherwise be lost.

In terms of practical application, pickled and fermented foods offer versatility in the prepper's diet. They can add flavor as well as nutrition to meals, even when fresh ingredients are unavailable. This variety is crucial for maintaining morale and well-being during extended periods of hardship. Furthermore, the skills gained through practicing these preservation methods empower individuals, providing an invaluable sense of control and competence in uncertain times.

In conclusion, the benefits of pickling and fermentation for preppers are manifold, encompassing food security, nutrition, safety, self-sufficiency, sustainability, cultural significance, and dietary diversity. These techniques embody the principles of preparedness, resilience, and adaptability, offering practical solutions for maintaining a stable as well as healthy food supply in the face of uncertainty. As preppers continue to hone their skills in these traditional preservation methods, they prepare themselves for potential future challenges and reconnect with cultural practices that have sustained humans for millennia. In embracing pickling and fermentation, preppers ensure that their preparation is about survival and maintaining a quality of life that values health, community, and sustainability.

## Overview of what the book will cover

This book is a comprehensive exploration into contemporary resilience and adaptability. It is designed to serve as both a guide and a source of inspiration for individuals desiring to navigate the complexities of the modern world. The book aims to cover various topics, each carefully chosen to equip the reader with knowledge, strategies, and insights to thrive in an ever-changing environment. This section provides an overview of the various subjects the book will explore, offering a glimpse into its depth and breadth.

At the heart of the book is an exploration of resilience, both at an individual and community level. It begins with a foundational understanding of what resilience means in the context of personal development, psychological well-being, and societal stability. The narrative then expands to illustrate how resilience can be cultivated through mindfulness practices, emotional intelligence, and the ability to adapt to adversity. Drawing on research from psychology, neuroscience, and anecdotal evidence, this section aims to guide readers towards developing a more resilient mindset.

Adaptability, a key theme of the book, is thoroughly examined in the context of rapid technological change, economic shifts, and environmental challenges. The book addresses how individuals and organizations can remain flexible and innovative amidst uncertainty. It discusses strategies for embracing change, the importance of continuous learning, and the necessity of being open to new ideas and methods. This section prepares the reader for the inevitable transformations that will occur in their personal and professional lives.

In addition, a sizeable amount of the book is devoted to discussing the impact that technology will have on our future. It critically analyzes emerging trends, such as artificial intelligence, blockchain, and renewable energy technologies, and their potential impacts on society. The discussion is balanced, highlighting both the opportunities these advancements present for improving quality of life and the ethical dilemmas and challenges they pose. This section aims to provoke thoughtful consideration on navigating the technological landscape responsibly and effectively.

Sustainability and environmental stewardship are also central themes, reflecting the growing concern for the planet's health and future generations' well-being. The book explores practical approaches to reducing one's carbon footprint, engaging in sustainable consumption, and supporting eco-friendly initiatives. Through showcasing success stories and innovative solutions, it inspires readers to take actionable steps towards a more sustainable lifestyle.

The book further delves into the importance of community and social connections in fostering resilience and adaptability. It examines how strong social networks can provide support during times of crisis, enhance mental health, and contribute to a sense of belonging and purpose. This section encourages readers to invest in their communities by volunteering, participating in local governance, or simply building stronger relationships with neighbors and colleagues.

In addressing the challenges of the 21st century, the book also explores themes of leadership, ethical decision-making, and the significance of diversity and inclusion. It offers insights into leading with empathy, fostering a culture of respect and understanding, and navigating ethical dilemmas in an increasingly complex world. This section is particularly relevant for those in positions of influence, whether in their organizations, communities, or broader society.

Personal well-being, a thread throughout the book, emphasizes the significance of keeping physical and mental health as foundational to resilience and adaptability. It covers topics such as stress management, the benefits of exercise and nutrition, and the importance of sleep and rest. Practical advice and tips are provided to help readers incorporate healthy habits into their daily lives, with the understanding that personal well-being is crucial for facing the demands of the modern world.

Finally, the book concludes with a forward-looking perspective, urging readers to envision a future characterized by resilience, adaptability, and sustainability. It calls for a collective effort to solve global challenges, emphasizing the power of individual actions in contributing to a larger change. This concluding section serves as a call to action, inspiring readers to apply the principles discussed in the book to their lives and communities.

In conclusion, this book promises to be an invaluable resource for those desiring to understand and navigate the complexities of the contemporary world. Covering a diverse range of topics—from personal development and technological advancements to environmental sustainability and community building—aims to provide readers with the tools and insights needed for a resilient and adaptable life. Through a combination of theoretical frameworks, practical advice, and real-world examples, the book seeks to empower individuals to thrive in an ever-changing landscape, fostering a future that is not only sustainable but also enriched by diversity, innovation, and connectedness.

# CHAPTER II

# Understanding the Science Behind Pickling and Fermentation

### Explanation of microbial fermentation

Microbial fermentation is a fascinating and complex biological process that has been utilized for thousands of years, albeit not always understood in the terms we know today. It involves using microorganisms, such as bacteria, yeasts, and molds, to convert organic substrates into desirable products, including foods, beverages, pharmaceuticals, and fuels. This section delves into the science of microbial fermentation, its mechanisms, applications, and its significant impact on various industries and aspects of human life.

At its core, microbial fermentation is a metabolic process that occurs in an anaerobic environment, meaning it takes place in the absence of oxygen. Microorganisms break down organic compounds to obtain energy, producing various end products such as ethanol, lactic acid, carbon dioxide, and other metabolites in the process. The specific outcome of fermentation depends on the type of microorganism involved and the substrates they metabolize. This biochemical process is essential for the survival of these microorganisms and has been harnessed by humans for various applications.

One of the most ancient and well-known microbial fermentation applications is in producing alcoholic beverages. Yeasts, particularly Saccharomyces cerevisiae, convert sugars found in fruits, grains, and other carbohydrate sources into ethanol and carbon dioxide. This process is the foundation of beer, wine, and spirit production, each beverage requiring specific conditions and yeast strains to achieve desired flavors and alcohol content. Similarly, lactic acid fermentation, primarily involving Lactobacillus species, is crucial in the production of fermented dairy products which includes yogurt and cheese, as well as fermented vegetables like sauerkraut and kimchi. Here, sugars are converted into lactic acid, which serves as a natural preservative and imparts a characteristic sour taste to these foods.

Beyond food and beverage production, microbial fermentation plays a vital role in the pharmaceutical industry. It is a critical method in producing antibiotics, hormones, and vaccines. For instance, the antibiotic penicillin is produced through the fermentation of the mold Penicillium chrysogenum. Recombinant DNA technology has further expanded the possibilities of microbial fermentation, allowing for the production of human proteins, such as insulin and the growth hormones, by genetically modified microorganisms. This has revolutionized the treatment of diseases like diabetes and dwarfism, making these essential proteins more accessible and reducing the risk of allergic reactions associated with animal-derived products.

Microbial fermentation also has significant applications in the biofuel industry, particularly in producing bioethanol and biobutanol from agricultural waste and other biomass. This sustainable energy source offers a promising alternative to fossil fuels, minimizing greenhouse gas emissions and dependence on non-renewable energy sources. Additionally, fermentation processes contribute to waste management and sustainability by converting organic waste into valuable products, including biogas, which can be used for heating and electricity generation.

Microbial fermentation mechanisms involve a series of enzymatic reactions that break down intricate organic molecules into simpler ones. These reactions are particular to the microorganism and the substrate involved. Enzymes, which are proteins that act as catalysts in biological reactions, play a crucial role in determining the efficiency and direction of fermentation processes. Understanding these mechanisms is essential for optimizing fermentation conditions, improving yields, and developing new products.

The control and optimization of fermentation processes are critical for industrial applications. Parameters such as temperature, pH, oxygen levels, and nutrient availability must be carefully regulated to ensure the desired outcome. Developing fermentation technology, including bioreactors and fermentation tanks designed to maintain optimal conditions, has dramatically enhanced the efficiency and scalability of microbial fermentation processes.

In conclusion, microbial fermentation is a cornerstone of biotechnology, with profound implications for food production, medicine, environmental sustainability, and energy. This ancient process, once a mystery, is now understood to be a complex interplay of biochemistry and microbial ecology. It exemplifies the incredible potential of harnessing natural processes for human benefit. As research continues to unveil new applications and improve existing ones, microbial fermentation is a testament to the symbiotic relationship between humans and the microscopic world, offering endless possibilities for innovation and sustainability. The exploration and application of microbial fermentation are bound to continue evolving, driven by scientific curiosity and the ever-present need to address global challenges through biotechnological advancements.

## Role of beneficial bacteria and yeasts

The role of beneficial bacteria and yeasts in pickling and fermentation is a fascinating topic that sits at the intersection of biology, gastronomy, and culture. These microorganisms transform raw ingredients into products that are not only preserved for longer periods but also enhanced in flavor, nutritional value, and digestibility. This section explores the intricate ways in which beneficial bacteria and yeasts contribute to the pickling and fermentation processes, highlighting their significance in culinary practices, food preservation, and human health.

At the core of pickling and fermentation is the action of beneficial microorganisms, primarily certain types of bacteria and yeasts, which metabolize the natural sugars present in foods, producing a range of by-products including acids, alcohol, and gases. This metabolic activity fundamentally transforms the food, resulting in products that are distinct from their original forms in taste and texture. In the case of pickling, the process can be initiated either through adding an acidic solution, typically vinegar, or through fermentation by natural or added bacteria that produce lactic acid, thereby lowering the pH of the environment and preserving the food.

Lactic acid bacteria (LAB) play a pivotal role in many fermentation processes, including the making of yogurt, sauerkraut, kimchi, and sourdough bread. These bacteria thrive in anaerobic (oxygen-free) conditions and efficiently convert sugars into lactic acid. The acidification of the environment inhibits the growth of spoilage-causing and pathogenic microorganisms, effectively preserving the food. Furthermore, LAB contributes to developing fermented foods' distinctive sour flavor characteristic. This group of bacteria includes various species and strains, each contributing uniquely to the fermented product's flavor, texture, and nutritional profile.

Yeasts, another group of microorganisms crucial to certain types of fermentation, are primarily involved in producing alcoholic beverages and bread. In the fermentation of beer, wine, and spirits, yeast metabolizes sugars to produce alcohol and carbon dioxide. The type of yeast used, along with fermentation conditions, influences the flavor and quality of the beverage. In bread making, the carbon dioxide produced by yeast is responsible for leavening the dough, creating the light and airy texture of baked bread. Additionally, yeasts contribute to developing complex flavors by producing various aromatic compounds.

The beneficial effects of bacteria and yeasts in fermentation extend beyond preservation and flavor enhancement to include significant health benefits. Fermented foods are known for their probiotic properties, with LAB, in particular, being recognized for their positive impact on gut health. These microorganisms can help balance the gut microbiota, support the immune system, and improve digestion and absorption of nutrients. The fermentation process also increases the bioavailability of certain nutrients, making fermented foods a valuable addition to the diet.

Moreover, the action of these microorganisms during fermentation can reduce the levels of certain anti-nutrients, such as phytates, which bind to minerals and hinder their absorption. By breaking down these compounds, fermentation makes minerals like iron, zinc, and calcium more available to the body. Additionally, the fermentation process can produce beneficial compounds, including vitamins (notably B vitamins), antioxidants, and bioactive peptides, further enhancing the nutritional value of the food.

Culturally, beneficial bacteria and yeasts used in pickling and fermentation have deep roots in human history, with evidence of fermented foods and beverages dating back thousands of years. These practices have been integral to human survival, allowing food storage over lean periods and long journeys. Today, the appreciation for fermented foods extends beyond their practical benefits to include an interest in their unique flavors, health attributes, and the role they play in cultural identity and culinary diversity.

In conclusion, beneficial bacteria and yeasts are indispensable to the processes of pickling and fermentation, transforming simple ingredients into complex, flavorful, and nutritious foods. These microorganisms preserve food and enhance its taste, texture, and health benefits through their metabolic activities. The continued study and application of these traditional food processing techniques, with an understanding of the underlying microbial processes, offer exciting possibilities for innovation in food science, nutrition, and gastronomy. As we gain a more profound appreciation for the role of these microorganisms, we can better leverage their potential to enrich our diets, support our health, and sustain our cultural heritage.

## Chemical processes involved in pickling and fermentation

The chemical processes involved in pickling and fermentation are complex, fascinating phenomena that have been utilized for thousands of years to preserve food, enhance its flavors, and improve its nutritional value. These methods, deeply rooted in human culture and tradition, involve a series of biochemical reactions that transform food substrates into preserved products with distinct tastes, textures, and health benefits. This section delves into the intricate chemical processes underpinning pickling and fermentation, exploring the roles of various substances, microorganisms, and environmental conditions in these transformative practices.

Pickling is a preservation method that entails immersing food items in an acidic solution, typically vinegar, or a saltwater brine, creating an environment where bacteria conducive to spoilage cannot thrive. The primary chemical process in pickling with vinegar involves the penetration of acetic acid into the food, which lowers its pH and inhibits the growth of harmful microorganisms. Acetic acid, a potent antimicrobial agent, denatures proteins and enzymes that spoilage bacteria need to survive and multiply. In brine pickling, or lacto-fermentation, salt plays a crucial role by drawing water out of the food through osmosis, creating a brine where specific beneficial bacteria can flourish.

Fermentation, on the other hand, is a metabolic process that happens in the absence of oxygen (anaerobic conditions), where microorganisms such as bacteria, yeasts, and molds convert organic compounds—primarily carbohydrates like sugars and starches—into alcohol, gases, or organic acids. The most common form of fermentation relevant to food preservation and enhancement is lactic acid fermentation. Lactic acid bacteria (LAB) are responsible for this process, which involves the conversion of carbohydrates found in the food into lactic acid. A decrease in the pH of the food is brought about by the build-up of lactic acid, which results in the formation of an acidic environment that is hostile to harmful pathogens but advantageous for the bacteria that are responsible for fermentation. This environment helps to preserve the food and contributes to its distinctive sour flavor.

Lactic acid fermentation involves several key steps. Initially, enzymes break down the complex carbohydrates in food into simpler sugars like glucose, which are more readily fermentable. The LAB then utilizes these sugars in a process known as glycolysis, where the sugar molecules are broken down into pyruvate. The pyruvate is subsequently converted into lactic acid, carbon dioxide, and other compounds, depending on the specific bacteria involved and the fermentation conditions. This biochemical transformation preserves the food and enriches it with probiotics—beneficial bacteria that contribute to gut health—vitamins, and enhanced flavors.

Another critical aspect of the fermentation process is the role of yeasts, particularly in producing alcoholic beverages and leavened bread. Yeasts, such as Saccharomyces cerevisiae, ferment sugars into ethanol and carbon dioxide through alcoholic fermentation. This process is similar to lactic acid fermentation in its initial stages but differs in the final products. The carbon dioxide generated by yeast fermentation is what causes bread to rise, while the ethanol contributes to the alcohol content in beverages like beer and wine.

The environment in which pickling and fermentation occur is crucial to the success of these processes. Factors such as temperature, pH, salinity, and the presence of oxygen can significantly affect the growth and activity of the microorganisms involved. For instance, lactic acid fermentation typically requires anaerobic conditions, moderate temperatures, and a specific range of salt concentration to favor the growth of LAB over harmful bacteria. Similarly, the acidity level (pH) in pickling must be carefully controlled to ensure that the food is preserved effectively without compromising its texture or nutritional quality.

In conclusion, the chemical processes involved in pickling and fermentation are intricate and multifaceted, involving a delicate balance of microorganisms, substrates, and environmental conditions. Through the action of acids, salts, and microbial metabolism, these methods transform food to enhance its longevity, nutritional profile, and sensory attributes. The scientific understanding of these processes sheds light on the mechanisms behind traditional food preservation techniques and opens up avenues for innovation in food science, improving the safety, quality, and health benefits of fermented and pickled foods. As we explore the complexities of these biochemical transformations, we gain a more profound appreciation for the art and science of food preservation that has sustained human civilizations throughout history.

## Factors affecting fermentation (temperature, salt concentration, pH)

Fermentation is a biological process that has been harnessed throughout human history to preserve food, enhance flavors, and produce alcoholic beverages. This process, fundamentally reliant on the metabolic activities of microorganisms such as bacteria and yeast, is influenced by several environmental factors. Temperature, salt concentration, and pH are paramount in determining the rate, efficiency, and safety of fermentation. Understanding how these factors affect fermentation is crucial for both traditional and industrial practices, ensuring the desired outcome of this intricate biochemical process.

Temperature plays a pivotal role in fermentation, as it directly affects the metabolic rate of fermenting microorganisms. Each bacteria or yeast species has an optimal temperature range that supports its growth and activity. For instance, the lactic acid bacteria involved in sauerkraut and kimchi fermentation thrive at temperatures between 20°C and 25°C. At temperatures below this range, the metabolic activity of the bacteria slows down, leading to longer fermentation times. Conversely, temperatures above this range can inhibit bacterial growth or promote the activity of undesirable microorganisms that could spoil the food. Temperature control is thus essential for achieving successful fermentation, as it ensures the dominance of beneficial microbes while suppressing harmful ones.

Salt concentration is another critical factor influencing fermentation. Salt acts as a selective agent, creating an environment that favors the growth of certain microorganisms while inhibiting others. In lactic acid fermentation, a moderate salt concentration promotes lactic acid bacteria growth by drawing water out of the cells of both the food and competing microbes through osmosis, effectively dehydrating and inhibiting the growth of spoilage organisms. However, too high a salt concentration can also impede the development of the beneficial lactic acid bacteria, slowing down the fermentation process or stopping it altogether. The appropriate salt concentration varies depending on the type of fermentation and the desired end product but typically ranges from 2% to 5% by weight.

The pH level, or the measure of acidity, is a crucial factor in fermentation that affects microbial growth and the safety of the fermented product. Most fermentative microorganisms, including lactic acid bacteria, thrive in mildly acidic environments. As these bacteria ferment sugars into lactic acid, the accumulating acid lowers the pH of the fermenting mixture. This drop in pH further inhibits the development of harmful bacteria, such as those that cause foodborne illnesses, making acidic fermented foods like yogurt, kefir, and pickles inherently safe. The critical pH for food safety in fermentation is generally considered to be 4.6 or below, under which most pathogenic bacteria cannot survive. Monitoring and adjusting the pH is thus vital not only for the progression of fermentation but also for ensuring the microbial safety of the final product.

These three factors—temperature, salt concentration, and pH—are interrelated and collectively determine the environment in which fermentation occurs. Manipulating these conditions allows for the control of microbial activity, guiding the fermentation process towards the desired outcome, whether it be the production of a specific flavor, texture, or nutritional profile. For example, adjusting the salt concentration and temperature can influence the speed of fermentation and the development of specific flavors by promoting the growth of certain strains of bacteria or yeast.

Moreover, the significance of these factors extends beyond the practical aspects of fermentation to impact the health benefits of the final product. Fermented foods are known for their probiotic properties, contributing to gut health and overall wellness. The balance of temperature, salt concentration, and pH during fermentation ensures the proliferation of beneficial microbes and enhances the bioavailability of nutrients, making fermented foods a valuable addition to the diet.

In conclusion, temperature, salt concentration, and pH are fundamental factors affecting fermentation. Their careful management is essential for optimizing beneficial microorganisms' growth and activity, ensuring fermented products' safety and quality. Whether in traditional home fermentations or industrial-scale production, understanding the role of these environmental conditions is critical to harnessing the full potential of fermentation. Through this knowledge, we can continue to enjoy and innovate within fermentation's rich culinary and health traditions, exploring new flavors, textures, and nutritional benefits.

# CHAPTER III

# Essential Equipment and Ingredients

### Tools needed for pickling and fermentation

Embarking on the journey of pickling and fermentation introduces one to a world where the ancient melds with the modern, where time-honored traditions meet contemporary culinary experimentation. Both pickling and fermentation are arts in their own right, each requiring a unique set of tools to transform raw ingredients into preserved delights. This section explores the essential tools needed for pickling and fermentation, highlighting how each item contributes to the process, ensuring success, safety, and flavor in every batch.

At the core of both pickling and fermentation are containers, which serve as the primary vessels for the transformation of ingredients. Glass jars, particularly wide-mouth mason jars, are favored for their non-reactive nature, ensuring that no unwanted flavors are imparted to the pickles or fermented foods. These jars also allow for easy monitoring of the fermentation process, offering a clear view of the developing colors and bubbles that indicate active fermentation. For larger batches, food-grade plastic buckets or ceramic crocks can be used, especially for traditional fermentations like sauerkraut or kimchi, where larger volumes are often prepared.

Lids and airlocks are crucial for creating the anaerobic (oxygen-free) environments necessary for fermentation. While pickling might not always require such an environment, fermentations do, to prevent the growth of mold and spoilage bacteria. Screw-on plastic lids with built-in airlocks or water-sealed fermentation crocks are popular choices. These tools allow carbon dioxide, a byproduct of fermentation, to escape while avoiding outside air from entering, thereby protecting the fermenting food from contamination.

Weights play a significant role in fermentation, mainly when whole vegetables or fruits are being fermented. They keep the produce submerged in brine, ensuring an even and safe fermentation process by preventing exposure to air, which could lead to spoilage. Glass weights, ceramic plates, or even clean, boiled rocks can serve this purpose, provided they are non-reactive and can be sanitized.

The importance of measuring tools cannot be overstated in pickling and fermentation. Accurate measurements ensure the correct balance of salt, water, and other ingredients, which is critical for creating the ideal conditions for fermentation or pickling. Measuring cups and spoons, a kitchen scale for weighing ingredients, and a pH meter or test strips for monitoring acidity levels are indispensable tools for both beginners and experienced practitioners alike. They help maintain consistency and safety across batches, allowing for the replication of successful recipes and the adjustment of those that need improvement.

Knives and cutting boards are basic yet essential kitchen tools that facilitate the preparation of ingredients for pickling and fermentation. A sharp knife is necessary for chopping or slicing fruits and vegetables uniformly, ensuring they pickle or ferment at an even rate. Non-reactive cutting boards, such as those made of plastic or bamboo, are preferred to avoid cross-contamination and absorb unwanted flavors.

For pickling, pots and pans made of stainless steel or enameled cookware are essential for heating pickling brines. The non-reactive materials ensure that the vinegar or any acidic ingredients do not react with the cookware, which could lead to off-flavors or the degradation of the cookware itself. These pots are used to bring the vinegar, water, salt, and any pickling spices to a boil before pouring over the prepared produce, initiating the pickling process.

Cleaning and sterilization equipment is paramount in pickling and fermentation to prevent the introduction of harmful bacteria or yeasts. This includes bottle brushes for cleaning jars, large pots for boiling and sterilizing equipment, and solutions like diluted bleach or vinegar for sanitizing. Ensuring that all tools and containers are thoroughly cleaned and sterilized before use is a fundamental step in protecting the quality as well as safety of the final product.

Lastly, the role of specialized fermentation equipment, such as fermenting lids with Bluetooth connectivity for tracking fermentation stages or vacuum sealers for long-term storage, is becoming increasingly prominent in modern pickling and fermentation practices. While not essential, these tools offer convenience and precision, appealing to those looking to delve deeper into the science and art of these processes.

In conclusion, the tools needed for pickling and fermentation are as varied as the techniques themselves, ranging from simple knives and jars to sophisticated airlocks and pH meters. Each tool plays a specific role in ensuring the successful preservation of foods through these ancient methods, adapted to fit modern kitchens and culinary aspirations. Whether one is a novice or seasoned in the arts of pickling and fermentation, having the right tools on hand is the first step in a journey filled with flavorful, nutritious, and satisfying creations.

## Types of containers for fermentation

Fermentation, a process steeped in history and tradition, has been a cornerstone of culinary practices across cultures for centuries. Central to the success of this biological transformation is the choice of container in which it occurs. The type of vessel used can significantly influence the environment for the fermenting microorganisms, ultimately affecting the final product's flavor, texture, and safety. This section delves into the various types of containers suitable for fermentation, highlighting their characteristics, benefits, and considerations.

Glass jars are among the most popular choices for small-scale fermentation projects. Transparent and non-reactive, glass allows for monitoring the fermentation process without the risk of chemical interactions with the food. Wide-mouthed glass jars are particularly favored for their ease of access and cleaning. Whether sealing with a traditional lid or employing airlock systems designed to release gases while preventing outside air from entering, glass jars offer versatility for a range of fermentations, from sauerkraut to kimchi and pickles.

Ceramic crocks, often associated with traditional fermentation practices, are prized for their aesthetic appeal and functionality. Thick-walled and sturdy, these vessels provide excellent insulation against temperature fluctuations, maintaining a stable environment that is conducive to steady fermentation. Many ceramic crocks come equipped with weights to submerge the ferment in brine, a critical factor in anaerobic fermentation processes. However, when opting for a ceramic container, one must ensure it is lead-free and glazed with a food-safe coating to avoid contamination.

Plastic containers, particularly those made from food-grade plastics such as HDPE (high-density polyethylene) or PET (polyethylene terephthalate), present a lightweight, break-resistant option. Their affordability and durability make plastic vessels appealing for beginners and those fermenting in large quantities. Nevertheless, caution is advised when using plastic for fermentation; it should be free from BPA (bisphenol A) and other harmful chemicals that could leach into the food. Additionally, plastic is more prone to scratching, which can harbor unwanted bacteria, thus requiring meticulous cleaning and maintenance.

Stainless steel containers offer another viable option for fermentation, known for their strength, durability, and ease of cleaning. Stainless steel is corrosion-resistant and does not react with acidic foods, making it suitable for various fermentation projects. However, it is essential to use high-quality, food-grade stainless steel to avoid the risk of metal leaching. While stainless steel cannot be used for all types of fermentation (such as those producing strong acids, which could corrode the material), it is excellent for fermenting beverages like beer and wine.

Wooden barrels have a long history in the fermentation of alcoholic beverages, which include wine and beer, and continue to be used for their unique ability to impart flavors to the ferment. The porous nature of wood allows for a small amount of oxygen exchange, which can benefit certain types of fermentation by promoting the development of complex flavor profiles. However, wooden vessels require rigorous maintenance to prevent contamination and spoilage, including regular cleaning and sterilization. The choice of wood, typically oak, is also crucial, as it can significantly influence the taste and character of the final product.

Earthenware pots, used in various cultures worldwide for centuries, provide a naturally insulating container for fermentation. Like ceramic crocks, earthenware is breathable, allowing for a slight exchange of gases that can benefit certain ferments. These pots often come with their traditional methods and recipes, such as the Korean onggi used for making kimchi. As with ceramic, it is essential to ensure that any earthenware used for fermentation is glazed with a food-safe material to prevent contamination.

Several factors must be considered in choosing the right container for fermentation, including the scale of the project, the specific requirements of the ferment, and the material's impact on the process and final product. Regardless of the type, the container must be clean and free from contaminants to ensure a successful and safe fermentation. Each type of vessel carries its history, characteristics, and unique set of benefits that can enhance the fermentation process, contributing to the richness of flavors and traditions that define fermented foods worldwide.

In conclusion, selecting an appropriate container is a critical step in the fermentation process, influencing not just the practical aspects of food preservation but also the cultural and sensory qualities of the fermented product. From glass jars and ceramic crocks to plastic buckets, stainless steel pots, wooden barrels, and earthenware, the diversity of containers reflects the versatility and global heritage of fermentation practices. Each material offers distinct advantages and considerations, allowing fermenters to experiment and discover the best fit for their culinary creations, preferences, and traditions.

## Essential ingredients (vegetables, salt, water, spices)

The art of pickling and fermentation has been a staple in culinary traditions worldwide for centuries, preserving the bounty of harvests and enriching diets with flavorful, nutritious foods. At the heart of these processes are a few essential ingredients—vegetables, salt, water, and spices—that together work to transform fresh produce into something entirely new and delightful. Understanding how each component contributes to the pickling and fermentation processes not only sheds light on the science behind these methods but also highlights the simplicity and elegance of preserving foods.

Vegetables are the foundation of pickling and fermentation. Virtually any vegetable can be pickled or fermented, with each type bringing its unique flavors, textures, and nutritional profiles to the finished product. Cucumbers, carrots, cabbages, and beets are among the most commonly used, but the possibilities are as vast as the variety of edible plants. In fermentation, particularly, the natural sugars present in vegetables serve as food for beneficial bacteria, initiating the fermentation process. These microorganisms, in turn, produce lactic acid, which acts as a natural preservative, enhancing the vegetable's shelf life and flavor. The choice of vegetables can significantly influence the final taste and texture of the product, making the selection a critical step in preparing pickled and fermented foods.

Salt is another indispensable ingredient in both pickling and fermentation. In pickling, salt, combined with water, creates a brine that submerges the vegetables, protecting them from spoilage by creating an environment hostile to harmful bacteria. Salt plays a pivotal role in fermentation by drawing water out of the vegetables, creating a brine in which lactic acid bacteria can thrive while inhibiting the growth of spoilage-causing microbes. The salt concentration is crucial; too much can inhibit fermentation, while too little may allow unwanted bacteria to proliferate. The type of salt used is also essential, as iodized salt or salt with anti-caking agents can interfere with the fermentation process. Thus, pure, non-iodized salt, such as sea salt or kosher salt, is recommended for both pickling and fermentation.

When used in pickling, water acts as a medium for the salt to dissolve into a brine, enveloping the vegetables and facilitating the preservation process. The water must be pure and free from impurities that could affect the pickling process, such as chlorine commonly found in tap water, which can inhibit the development of beneficial bacteria during fermentation. As such, filtered or distilled water is often preferred to ensure that the only microorganisms at work are those intended to ferment the vegetables. In fermentations where water is added, such as in making brine for sauerkraut or kimchi, its role is to create an anaerobic environment, crucial for the fermentation process to proceed correctly.

Spices and seasonings are the final touches that transform pickled and fermented vegetables from simple preserved foods into complex, flavor-rich delicacies.

Spices such as dill, mustard seeds, garlic, and chili peppers are common in pickling, each adding their distinct flavors to the finished product. In fermentation, spices not only contribute to the taste but can also offer additional antimicrobial properties, further ensuring the safety and success of the fermentation. The combination of spices used in pickling and fermentation is limited only by imagination and taste preferences, allowing for endless variations and the creation of unique, customized flavors.

When brought together under the right conditions, these essential ingredients undergo a remarkable transformation. The vegetables soften slightly yet retain a satisfying crunch, imbued with the flavors of the spices and the tangy depth created by the fermentation or pickling process. While crucial for preservation, the salt also enhances the natural flavors of the vegetables as well as spices, creating a harmonious balance that makes pickled and fermented foods such irresistible additions to meals.

In conclusion, the simplicity of the ingredients used in pickling and fermentation belies the complexity of flavors and textures they can achieve. Vegetables, salt, water, and spices, each with its role in the preservation process, come together to create more than just preserved foods—they are transformed. These ingredients do not just extend the shelf life of seasonal produce; they enhance nutritional value, introduce beneficial probiotics into the diet, and provide a canvas for culinary creativity. Through the careful selection and combination of these essential components, the ancient practices of pickling and fermentation continue to offer delicious, healthful foods that connect us to past and present traditions across cultures and cuisines.

## Safety precautions and sanitation practices

The processes of pickling and fermentation have been cherished methods of food preservation for centuries, allowing cultures around the world to extend the shelf life of their food and enjoy unique flavors. While these methods are both economical and sustainable, ensuring the safety of the final product requires meticulous attention to sanitation practices and safety precautions. This section delves into the crucial aspects of maintaining safety and cleanliness during pickling and fermentation, highlighting the importance of these practices in preventing foodborne illnesses and ensuring the production of high-quality, safe-to-eat foods.

Sanitation and cleanliness are the cornerstones of safe pickling and fermentation processes. Before beginning, it is essential to thoroughly clean and sterilize all equipment, containers, and surfaces that will come into contact with the food. This includes jars, lids, fermentation weights, cutting boards, and utensils. Proper sterilization can be achieved through boiling equipment for at least 10 minutes or using a solution of bleach and water, followed by thorough rinsing with boiled or distilled water to remove any residual chlorine. This step is critical in eliminating unwanted bacteria, yeasts, or molds that might spoil the food or pose health risks.

Choosing the right ingredients is also pivotal to the safety of pickled and fermented foods. Fresh, high-quality vegetables free from bruises or signs of spoilage are essential, as compromised produce can introduce harmful microorganisms into the fermentation environment. Similarly, using non-iodized salt without anti-caking agents and filtered or distilled water can help ensure that the fermentation process proceeds without interference from impurities that may inhibit the growth of beneficial bacteria.

Maintaining an appropriate environment for fermentation is crucial for the safety and success of the process. This involves monitoring temperature, as most pickling and fermentation processes require specific temperature ranges to encourage the growth of helpful microorganisms while inhibiting harmful ones. For example, sauerkraut and other vegetable ferments typically require a room temperature of about 18°C to 22°C (about 65°F to 72°F) for optimal fermentation. Too high temperatures can foster the growth of undesirable bacteria, while too low temperatures can slow down or halt the fermentation process.

Using appropriate salt concentrations in brine solutions is a key safety measure in pickling and fermentation. Salt serves as a natural preservative, inhibiting the growth of pathogenic bacteria. However, the salt concentration must be carefully balanced; too much can prevent the fermentation process altogether, while too little may not adequately suppress harmful microorganisms. Recipes from reputable sources or tested home preservation guides should be followed to ensure the salt concentration is effective and safe.

During fermentation, it is vital to keep the produce submerged under the brine to create an anaerobic (oxygen-free) environment, which is necessary for lactic acid bacteria to thrive and prevent mold and yeast growth on the surface. Using fermentation weights or a clean, boiled stone can help keep vegetables below the surface. If mold does appear on the surface, it is often a sign that the ferment has been exposed to air or that the salt concentration was insufficient. While some sources suggest removing the mold and consuming the rest, it is safest to discard the entire batch to avoid possible health risks.

Monitoring the pH of the ferment is another essential safety practice. Lactic acid production during fermentation naturally lowers the pH, creating an acidic environment hostile to pathogenic bacteria. A pH of 4.6 or lower is generally considered safe for preventing the growth of botulinum bacteria. pH strips or digital pH meters can be used to monitor acidity levels throughout the fermentation process. The product should not be consumed if the desired pH is not reached.

Finally, correctly storing pickled and fermented products is essential for maintaining their safety and quality. Refrigeration after fermentation can slow down microbial activity, preserving the product's flavor and texture while keeping it safe to eat. To prevent contamination and spoilage, fermented foods should be stored in clean, airtight containers.

In conclusion, while pickling and fermentation are effective and natural methods of food preservation, their safety relies heavily on strict sanitation practices and adherence to proper techniques. One can minimize the risk of foodborne illnesses by thoroughly cleaning and sterilizing equipment, selecting quality ingredients, controlling environmental factors, and monitoring the fermentation process closely. These precautions ensure that the ancient art of pickling and fermentation remain safe, enjoyable, and beneficial in modern culinary endeavors.

# CHAPTER IV

# Pickling Techniques

## Basic pickling methods (vinegar-based)

Pickling, a method steeped in tradition and practicality, has been a cornerstone of food preservation for centuries. Among the various techniques employed, vinegar-based pickling stands out for its simplicity, versatility, and the delightful tang it imparts to a wide array of foods. This section explores the fundamental aspects of vinegar-based pickling, delving into the process, the science behind it, and the myriad ways in which it can be customized to suit different tastes and culinary traditions.

At its core, vinegar-based pickling involves submerging food items, typically vegetables or fruits, in a solution of vinegar, often combined with water, salt, and an assortment of spices and seasonings. The primary purpose of this method is preservation, with the acidic environment created by the vinegar serving to inhibit the growth of harmful bacteria that cause spoilage. However, beyond mere preservation, vinegar-based pickling also transforms the texture and flavor of the produce, resulting in a delicious and durable product.

The process begins with selecting the produce to be pickled. Virtually any vegetable and/or fruit can be pickled using the vinegar-based method, though some of the most popular choices include cucumbers, carrots, onions, and peppers. The selected items are cleaned and, if necessary, peeled or cut into desired shapes and sizes. The preparation of the produce is crucial, as it affects not only the pickles' appearance but also their texture and flavor absorption.

Next, the pickling liquid is prepared. While vinegar is the main ingredient, its type and concentration can vary widely, affecting the final product's flavor profile. White vinegar, with its clean, sharp taste, is commonly used for its neutrality, allowing the produce's and spices' flavors to shine through. Apple cider vinegar offers a fruitier undertone, while wine vinegars introduce a more complex flavor palette. The vinegar is usually diluted with water to moderate its intensity, with the ratio of vinegar to water affecting the pickle's acidity and taste.

Salt is another key component of the pickling liquid. It enhances flavor and inhibits microbial growth. Like vinegar, the type of salt used can vary, though non-iodized salts such as kosher or pickling salt are preferred to avoid clouding the pickling liquid. Spices and seasonings are added to the mix to infuse the pickles with desired flavors. Common choices include dill, garlic, mustard seed, and peppercorns, though the possibilities are limited only by the pickler's imagination.

The prepared produce is then packed into sterilized jars, and the hot pickling liquid is poured over them, ensuring that the items are completely submerged. This step is critical, as exposure to air can lead to spoilage. The jars are sealed and left to cool before being stored. While some pickles are ready to eat within hours, others may benefit from weeks or even months of aging, during which their flavors continue to develop and mature.

The science behind vinegar-based pickling lies in the acidity of the vinegar, which creates an environment that is inhospitable to most bacteria, including those that cause food spoilage. The acetic acid in vinegar is a potent antimicrobial agent, effectively preserving the produce while imparting the tangy flavor associated with pickled foods. Additionally, the osmotic pressure created by the salt draws water out of the cells of the produce, further inhibiting bacterial growth and contributing to the pickles' crunchy texture.

Vinegar-based pickling offers an immense scope for customization and creativity. Beyond the choice of vinegar, salt, and spices, variations in the preparation method can result in vastly different products. Quick pickles, made by pouring hot vinegar over the produce and refrigerating for a short period, offer a crisp, fresh-tasting alternative to traditional pickles. On the other end of the spectrum, long-process pickles, which involve aging the sealed jars for several weeks or months, develop deeper, more complex flavors.

In conclusion, vinegar-based pickling is a multifaceted technique that marries preservation with flavor enhancement. Its simplicity belies the depth of science and tradition behind it, offering a canvas for culinary experimentation. Whether aiming for the bright, crisp taste of quick pickles or the rich complexity of aged varieties, vinegar-based pickling provides a means to capture the essence of the harvest, extend the life of seasonal produce, and enjoy the tangy delights of pickled foods year-round. Through this age-old method, we preserve food and the rich tapestry of culinary heritage that pickling represents.

## Quick pickling vs. traditional pickling

Pickling, an age-old practice, has long been cherished for its ability to preserve food beyond its natural shelf life and imbue it with distinct flavors. At its core, pickling involves submerging foods in an acidic solution or through fermentation in brine, creating an environment inhospitable to spoilage-causing bacteria. The pickling techniques are diverse but broadly fall into two categories: quick pickling and traditional pickling. Each method presents a unique benefits and caters to different preferences and time constraints, illustrating the versatility and enduring popularity of pickling across cultures.

As the name suggests, quick pickling is a rapid method that allows for the consumption of pickled goods within a short timeframe, often within hours or days after preparation. This method involves submerging the vegetables or fruits in a combination of vinegar, water, salt, and sugar, frequently heated to dissolve the solids and allow the flavors to meld. Spices and herbs are added to the brine to infuse the pickles with desired flavors. The key characteristic of quick pickling is its speed and ease, making it an accessible option for home cooks looking to add a tangy crunch to their meals without the commitment required for traditional pickling methods. Quick pickles are typically stored in the refrigerator, where the cold temperature slows any further fermentation or spoilage, keeping them crisp and flavorful for a few weeks.

In contrast, traditional pickling is a slower process that relies on natural fermentation to acidify the food. This method often uses a saltwater brine, into which vegetables are submerged. Over time, the natural lactobacillus bacteria present on the surface of the vegetables begin to ferment the sugars into lactic acid, gradually lowering the pH of the brine and creating an acidic environment that preserves the vegetables.

Traditional pickling can take from several days to a few months, depending on the specific recipe and the desired level of sourness. This method extends the shelf life of the pickled items significantly and enhances their nutritional value, introducing beneficial probiotics that support gut health.

The differences between quick and traditional pickling extend beyond the time required for each process. Due to its short preparation time, quick pickling offers immediate gratification and the ability to experiment with a wide variety of flavors as well as ingredients. It is a great way to preserve seasonal produce or add a zesty component to meals without waiting for the lengthy fermentation process. However, quick pickles lack the probiotic benefits of traditionally fermented pickles, as the high vinegar content and refrigeration inhibit the growth of beneficial bacteria.

Traditional pickling, on the other hand, demands patience and a greater understanding of the fermentation process.

The flavors developed through traditional pickling are often more complex and nuanced than those achieved through quick pickling, with the gradual fermentation process allowing for a deeper melding of the brine's flavors with the vegetables. Additionally, the probiotic content of traditionally pickled foods contributes to their appeal, offering health benefits alongside their culinary uses. However, traditional pickling requires careful monitoring of environmental conditions, such as temperature and cleanliness, to ensure the fermentation process proceeds safely and desired outcomes are achieved.

Both quick and traditional pickling methods have their place in the culinary world, each offering distinct advantages that cater to different needs as well as preferences. Quick pickling provides a convenient and fast way to enjoy pickled vegetables, ideal for those looking to experiment with flavors or add a tangy element to their dishes without the wait. While more time-consuming, traditional pickling rewards patience with complex flavors and health benefits, preserving not just food but also a piece of cultural heritage passed down through generations.

In conclusion, the choice between quick and traditional pickling depends on the individual's goals, time constraints, and preferences. Whether seeking the immediate satisfaction and convenience of quick pickles or the depth of flavor and nutritional advantages of traditional fermented pickles, both methods enrich our diets and culinary traditions. They underscore the creativity and ingenuity of human food preservation techniques, allowing us to enjoy a bounty of flavors and nutrients well beyond the growing season.

# Recipes for pickled vegetables (cucumbers, carrots, beets)

The tradition of pickling vegetables, from cucumbers and carrots to beets, is a culinary practice that transcends cultures, offering a tangy and often spicy accompaniment to meals while preserving the garden's bounty. This section delves into the art of pickling these vegetables, providing recipes that encapsulate the essence of pickling by transforming these humble ingredients into delectable, preserved delights. Each recipe is tailored to highlight the unique flavors and textures of the vegetable, creating a variety of pickles that cater to different palates and prcfcrences.

Starting with cucumbers, the quintessential pickled vegetable, we explore a classic dill pickle recipe. This recipe begins by preparing the cucumbers, which are best when fresh, crisp, and ideally of a smaller variety such as Kirby or Persian, known for their firmness and thin skins. The cucumbers are washed and then cut into the desired shapes—slices, spears, or left whole for a traditional presentation. The brine, a crucial component of any pickle, combines water, white vinegar, and salt, brought to a simmer until the salt dissolves. Fresh dill, garlic cloves, and mustard seeds are added to sterile jars before tightly packing in the cucumbers. The hot brine is then poured over the cucumbers, completely submerging them. The jars are sealed and left to cool at room temperature before being transferred to the refrigerator. After a minimum of 48 hours, the cucumbers transform into crunchy, flavorful dill pickles, their taste a perfect balance of tangy, salty, and herbaceous notes, making them a favorite for snacking or as a side dish.

With their natural sweetness and crisp texture, carrots offer a different pickling perspective. A pickled carrot recipe often includes a sweet and spicy brine to complement the carrots' flavor profile. The carrots are peeled and cut into sticks or rounds, depending on preference. The brine for carrots integrates apple cider vinegar with water, sugar, and salt, creating a sweet and tangy base. To this, spices such as coriander seeds, garlic, and a hint of chili flakes are added for warmth and depth. The carrots are blanched briefly to soften them slightly before being placed with spices in jars. The hot brine is then poured over the carrots, and the jars are sealed and cooled. After a few days in the refrigerator, these pickled carrots emerge as a vibrant, sweet, and slightly spicy condiment that pairs wonderfully with sandwiches, salads, or as part of a charcuterie board.

With their earthy sweetness and deep, rich color, beets are transformed through pickling into a visually stunning and deliciously complex treat. The preparation of pickled beets begins with roasting them to enhance their natural sugars. Once cooled, the skins are removed, and the beets are sliced or cubed. The brine for beets is a mixture of red wine vinegar and water, sweetened with sugar and enhanced with salt. Spices such as cloves, cinnamon, and allspice add a warm spice profile, complementing the beets' earthiness. The beets are placed in jars with the spices, covered with the brine, and then sealed. The beets are ready to enjoy after pickling in the refrigerator for at least a week. Their flavor is a complex blend of sweet, sour, and spice, making them an excellent addition to salads, a unique topping for burgers, or a standalone side dish.

These recipes for pickled cucumbers, carrots, and beets showcase the versatility and creativity of pickling. By preserving vegetables in a vinegar-based brine infused with an array of spices and herbs, each vegetable is transformed into a pickle with its own distinct flavor profile. The process, though straightforward, requires attention to detail, from preparing the vegetables and brine to sterilizing the jars, ensuring that the final product is not only delicious but safe to consume.

In conclusion, pickling is a culinary art that marries the simplicity of preserving food with the complexity of flavors that can be acquired through the process. The recipes provided for cucumbers, carrots, and beets are but a starting point, encouraging exploration and experimentation with different vegetables, spices, and vinegar types. Pickling not only extends the shelf life of seasonal produce but also offers a way to enjoy the tastes of the garden year-round, adding depth and zest to meals. Whether enjoyed as a snack, a condiment, or an integral part of a dish, pickled vegetables are a testament to this ancient preservation method's enduring appeal and culinary value.

## Tips for achieving desired texture and flavor

Achieving the desired texture and flavor in pickling is an art that balances science with culinary craft. Pickling, a method steeped in tradition, transforms fresh produce into tangy, flavorful delights, preserving their essence while imbuing them with a new character. This transformation is guided by factors including the choice of produce, brine composition, and pickling method, each playing a pivotal part in the final texture and taste of the pickled goods. This section delves into the nuanced techniques and considerations for crafting pickles with optimal flavor and texture, offering guidance for both novice and seasoned picklers.

The foundation of a successful pickle begins with the selection of produce. Freshness is paramount, as the quality of the starting materials directly influences the outcome. Vegetables and fruits should be ripe yet firm, free from bruises or signs of spoilage, which can affect both texture and flavor. For example, cucumbers for pickling should be crisp, with a uniform green color and no soft spots. Similarly, carrots should be firm and vibrant in color. The size and cut of the produce also matter; smaller, uniform pieces not only pickle more evenly but also tend to retain better texture.

The composition of the brine, a solution of vinegar, water, salt, and often sugar, is crucial in determining the pickle's flavor profile. The type of vinegar used can dramatically alter the taste; white vinegar offers a sharp, clean acidity, apple cider vinegar brings a fruity tang, and rice vinegar contributes a mild, sweet acidity. The vinegar-to-water ratio also affects the pickle's strength and sharpness. A higher vinegar concentration results in a more intense flavor and increased preservation qualities, while a higher water content can soften the flavor but may also impact the pickle's shelf life. Beyond its preservative qualities, salt enhances the natural flavors of the produce and should be used in a balanced manner to avoid overpowering the pickle. Non-iodized salt is preferred to prevent any chemical tastes.

Sugar, when added, moderates the acidity of the brine, lending a subtle sweetness that complements the natural flavors of the vegetables or fruits. It is possible to adjust the amount of sugar to suit one's individual preferences as well as the flavor that is desired in the end. Spices and herbs, from dill and garlic to mustard seeds and peppercorns, are the artisans' tools for infusing the brine with depth and complexity. The combination of spices can be tailored to match the produce, with warmer spices like cloves and cinnamon pairing well with sweet vegetables like beets, and fresh herbs like dill and tarragon complementing green vegetables like cucumbers and beans.

The pickling process itself, whether quick pickling or fermentative pickling, influences the texture and flavor of the final product. Quick pickling, involving the pouring of hot brine over the produce and refrigerating, offers a faster turnaround but typically results in a crunchier texture and brighter flavors. This method is ideal for those seeking a quick addition to meals, as the pickles are ready to eat within hours to days. In contrast, fermentative pickling relies on the natural lacto-fermentation process, where salt draws out the moisture from the produce, creating an environment where beneficial bacteria thrive. This slower process develops complex, nuanced flavors and can result in a softer texture, depending on the fermentation duration.

Specific techniques can be employed to maintain the desired texture, particularly for crunchier pickles. Using grape leaves, horseradish leaves, or black tea in the pickling jar can introduce tannins, which help preserve the firmness of the vegetables. Additionally, ensuring that the produce is fully submerged under the brine can prevent exposure to air, which can lead to softening. For fermentative pickles, controlling the fermentation environment is key; too warm temperatures can accelerate fermentation, leading to softer textures, while too cool temperatures can slow down the process, affecting flavor development.

Achieving the desired flavor and texture in pickling is a delicate dance of selecting the right produce, crafting a balanced brine, and choosing the appropriate pickling method. It requires attention to detail, from the cleanliness of the equipment to the precision in measuring ingredients. Experimentation is also a vital part of the process, as it allows for discovering personal preferences and developing unique recipes. Whether aiming for crisp, tangy cucumber pickles, sweet and spicy pickled carrots, or richly flavored fermented sauerkraut, the joy of pickling lies in the journey from fresh produce to flavorful, preserved treasures. Through practice and patience, anyone is able to master the art of pickling, creating jars of delight that capture the ingredients' essence and the maker's imagination.

# CHAPTER V

# Fermentation Basics

### Introduction to lacto-fermentation

Lacto-fermentation is a fascinating and ancient culinary practice that harnesses the natural process of fermentation to preserve and enhance the flavor of foods. This method, which various cultures around the globe have utilized for thousands of years, involves the fermentation of vegetables, fruits, and other foods using lactic acid bacteria. These bacteria, naturally present on the surface of all living things, particularly on plants and soil, convert sugars and starches in food into lactic acid. This section introduces lacto-fermentation, outlining its historical context, the process's science, health benefits, and significance in culinary traditions.

Historically, lacto-fermentation served as a crucial means of food preservation before the advent of refrigeration. Cultures worldwide discovered that they could extend the shelf life of perishable goods through fermentation, ensuring a supply of edible food throughout scarce months. From sauerkraut in Germany, and kimchi in Korea to pickled vegetables across Eastern Europe and Asia, lacto-fermentation has been integral to human diet and survival. These conventional practices have been passed down through generations, becoming a part of many communities' cultural identity and culinary heritage.

The science of lacto-fermentation is both complex and fascinating. The process begins when vegetables or fruits are submerged in a brine solution or left to ferment in their own juice, creating an anaerobic (oxygen-free) environment. Lactic acid bacteria, primarily of the genus Lactobacillus, thrive in this environment, feeding on the natural sugars present in the food. As they metabolize these sugars, they produce lactic acid, which acts as a natural preservative. The acidification of the environment lowers the pH, making it inhospitable for harmful bacteria and pathogens to survive, thereby preserving the food. This process also results in the creation of beneficial enzymes, B-vitamins, Omega-3 fatty acids, as well as various strains of probiotics, which contribute to the nutritional value of the fermented product.

One of the most compelling aspects of lacto-fermentation is its health benefits. The lactic acid and probiotics produced during fermentation have been shown to enhance digestion, boost the immune system, and increase nutrient absorption. These probiotics, or "good" bacteria, are essential in maintaining gut health, directly influencing overall wellness. Fermented foods are also high in enzymes that can help in the breakdown of food as well as absorption of nutrients, making them a valuable addition to any diet.

Moreover, lacto-fermentation offers a sustainable and eco-friendly method of food preservation. It requires no electricity and minimal resources, minimizing food waste by means of extending the shelf life of produce. This makes it an attractive option for those looking to live a more sustainable lifestyle, contributing to a reduction in the carbon footprint connected with food storage and transportation.

The impact of lacto-fermentation on culinary traditions cannot be overstated. It introduces a depth of flavor and complexity to foods that would be unachievable through other means. The fermentation process can transform simple ingredients into something unique, adding sourness, tang, and umami to enhance the overall taste experience. Chefs and home cooks alike experiment with lacto-fermentation to create innovative dishes and flavors, pushing the boundaries of traditional cuisine.

Lacto-fermentation also fosters a connection to our culinary past, offering a link to the traditional foodways of our ancestors. Engaging in this practice can be an enriching experience, providing a sense of accomplishment and a physical connection to history and culture. It invites an exploration of the rich tapestry of fermented foods globally, encouraging culinary diversity and the preservation of heritage foods.

In conclusion, lacto-fermentation is a remarkable process that merges the art of cooking with the science of microbiology, offering a window into the ingenuity of traditional food preservation methods. Its benefits extend beyond mere longevity, contributing to nutritional enhancement, culinary innovation, and environmental sustainability. As we continue to rediscover and embrace these ancient techniques, lacto-fermentation stands as a testament to the enduring wisdom of our culinary forebears, enriching our diets and cultures in myriad ways. Whether as a hobbyist, a culinary professional, or simply a curious eater, exploring the world of lacto-fermentation opens up a universe of flavors, health benefits, and connections to the natural world.

## Creating a fermentation brine

Creating a fermentation brine is an essential step in the process of lacto-fermentation, a method of food preservation that not only extends the shelf life of perishable items but also enhances their nutritional value and flavor profile. This practice, deeply rooted in culinary traditions across the globe, involves submerging vegetables or fruits in a liquid medium conducive to the growth of beneficial lactic acid bacteria. These microbes, naturally present on the surface of all plant-based foods, thrive in the brine, fermenting the sugars present in the food into lactic acid. This section explores the intricacies of creating a fermentation brine, including the choice of water and salt, the importance of salt concentration, the inclusion of spices and seasonings, and the conditions necessary for successful fermentation.

At its most basic, a fermentation brine is a water and salt solution. However, the quality of these two ingredients plays a pivotal role in the success of the fermentation process. The water used should be free of chlorine and other chemicals typically found in tap water, as these can inhibit the growth of beneficial bacteria. Filtered, spring, or distilled water is recommended to ensure the fermentation environment is as pure as possible. The choice of salt is equally important, with sea salt or kosher salt preferred over table salt, which often has iodine and anti-caking agents that can interfere with the fermentation process. The salt flavors the food and creates an environment that selectively encourages the growth of lactic acid bacteria while deterring harmful pathogens.

The concentration of salt in the brine is a vital aspect that must be considered carefully. Too little salt can result in spoilage, while too much can inhibit fermentation entirely. A general guideline is a salt concentration of 2-5% by weight, depending on the specific recipe and desired outcome. For most vegetables, a 3% brine, equivalent to about 1 tablespoon of salt in every 2 cups of water, balances safety and flavor. This concentration allows for the proliferation of lactic acid bacteria, ensuring a successful fermentation process.

Spices and seasonings can be added to the brine to impart additional flavors to the fermented food. Garlic, dill, mustard seeds, peppercorns, and bay leaves are common additions, but the possibilities are limited only by the imagination. These ingredients can be tailored to complement the specific vegetables being fermented, creating a wide array of tastes and textures. When adding spices to the brine, it's essential to consider their potential impact on the fermentation process. Some spices may have antimicrobial properties that could affect the balance of microorganisms in the brine.

Creating the perfect fermentation brine is as much an art as it is a science. Once the brine is prepared, the vegetables or fruits are submerged, ensuring that the liquid completely covers them. This anaerobic environment is crucial for lacto-fermentation to occur, as exposure to air can introduce unwanted molds and yeasts. A weight or a smaller jar can be used to keep the produce submerged beneath the brine's surface.

The conditions under which the fermentation takes place also influence the outcome. The temperature, for example, should be carefully controlled. A room temperature of around 60-70°F (15-21°C) is ideal for most lacto-fermentation processes, as it allows for a steady, moderate fermentation rate. Too warm, and the fermentation could proceed too quickly, resulting in soft, overly sour vegetables; too cool, and the process may stall before the desired level of fermentation is achieved. Additionally, the fermentation vessel should be covered with a cloth or a loose lid to allow gases produced during fermentation to escape while keeping out contaminants.

Over the course of several days to weeks, the brine will become cloudy, a sign that fermentation is actively occurring. Tasting the ferment at various stages can help determine when it has reached the desired flavor and acidity. Once fermentation is complete, the vegetables can be transferred to the refrigerator, which slows down microbial activity, preserving the texture and flavor of the fermented product.

In conclusion, creating a fermentation brine is a nuanced process that requires attention to detail and understanding the factors that influence lacto-fermentation. The choice and quality of water and salt, the salt concentration, the addition of spices and seasonings, and the conditions under which fermentation occurs all play critical roles in the success of the process. By mastering the art of brine-making, one can unlock the full potential of lacto-fermentation, preserving food in a way that enhances its nutritional value, flavor, and digestibility. Whether a novice or an experienced fermenter, the journey of creating a fermentation brine is a rewarding exploration of the intersection between science, tradition, and culinary innovation.

## Fermentation vessels and their uses

Fermentation, an age-old method of preserving and enhancing the flavor of food, relies not just on the process itself, but also significantly on the vessels used to contain and nurture the ferment. The choice of fermentation vessel can influence the quality, taste, and safety of the fermented product, making it a crucial aspect of the fermentation process. This section explores various fermentation vessels, their unique characteristics, and how they are utilized in different fermentation practices.

Historically, fermentation vessels were crafted from the materials readily available to a culture, leading to a diversity of containers across different regions and traditions. This tradition continues with a wide range of options available, from glass jars to ceramic crocks, wooden barrels, stainless steel tanks, and plastic buckets. Each material offers distinct advantages and considerations, influencing its use in fermenting specific foods.

Glass jars are among the most popular choices for home fermenters due to their non-reactive nature, which ensures that no unwanted flavors are imparted to the ferment. Glass is also impermeable to gases, preventing the entry of oxygen, which could spoil the ferment while allowing for easy observation of the fermentation process. This transparency is invaluable for monitoring the progress and health of the ferment, making glass ideal for small-batch fermentation of vegetables, kombucha, and kefir.

Ceramic crocks have been used for centuries, particularly for large-batch fermentation of vegetables, such as sauerkraut and kimchi. The porous nature of ceramic allows for some breathability while providing excellent insulation, maintaining a stable temperature throughout the fermentation process. High-quality ceramic crocks are glazed with a lead-free finish to ensure food safety. Many come equipped with weights to keep the ferment submerged and a water seal that allows gases to escape without letting air in, minimizing the risk of contamination.

Wooden barrels, traditionally used for fermenting wine, beer, and vinegar, offer unique characteristics. Wood is naturally porous, allowing for micro-oxygenation which can benefit certain ferments by promoting the development of complex flavors. The type of wood, most commonly oak, also contributes tannins and other compounds, adding depth to the flavor profile of the ferment. However, wooden vessels require meticulous maintenance to prevent contamination and are best suited for those with experience in fermentation.

Stainless steel tanks are favored in commercial fermentation settings for their durability, ease of cleaning, and corrosion resistance. Stainless steel is non-reactive and provides an excellent barrier against oxygen, making it suitable for fermenting an array of products, such as wine, beer, and dairy ferments like yogurt. Its use in home fermentation is growing, especially for those committed to large-scale or frequent fermenting projects.

Plastic buckets, specifically those made from food-grade plastic, offer fermenters an affordable and lightweight option. They are particularly useful for large batches where glass or ceramic would be impractical due to weight or cost. However, ensuring the plastic is free from BPA and other harmful chemicals that could leach into the ferment is crucial. Additionally, plastic can be prone to scratching, which may harbor unwanted bacteria, thus requiring careful handling and cleaning.

In addition to these traditional materials, modern innovations have introduced fermentation vessels designed to simplify the process and enhance safety. These include glass or plastic vessels with built-in airlocks to allow gases to escape while preventing air entry, and silicone fermentation lids that can be used on regular mason jars, automating the burping process necessary in some fermentations.

The choice of fermentation vessel is not merely a matter of preference but should be guided by the specific requirements of the ferment, the scale of production, and concerns about safety and quality. For example, small-scale fermenters and beginners might prefer the simplicity and visibility of glass jars, while those with more experience or the need to produce larger quantities may opt for ceramic crocks or stainless steel tanks.

Furthermore, the cultural and traditional context of the ferment may dictate the use of specific vessels. For instance, the use of wooden barrels for wine fermentation is not just a matter of flavor but also a nod to tradition and craftsmanship, while the use of onggi pots in Korean kimchi fermentation connects the practice to its cultural roots.

In conclusion, fermentation vessels play a pivotal role in the art and science of fermentation. They do more than just contain the ferment; they influence the environment in which microbial magic occurs, affecting the final product's flavor, texture, and safety. From the humble glass jar to the large stainless steel tank, the choice of vessel reflects the fermenter's goals, the scale of their endeavor, and often, a connection to the cultural heritage of the ferment itself. As the interest in fermentation continues to grow, the diversity and innovation in fermentation vessels promise to expand, offering fermenters new tools to explore and enjoy the ancient fermentation practice.

## Commonly fermented vegetables (sauerkraut, kimchi)

Fermentation, a culinary practice steeped in history, transforms simple vegetables into complex, flavor-rich foods that are both nutritious and long-lasting. Among the myriad of fermented foods enjoyed around the world, sauerkraut and kimchi stand out as iconic examples of how fermentation can elevate vegetables to staples of cultural cuisine. This section delves into the origins, preparation methods, nutritional benefits, and cultural significance of these commonly fermented vegetables, offering insight into their enduring popularity.

Sauerkraut, meaning "sour cabbage" in German, is a testament to the simplicity and effectiveness of lacto-fermentation. Its origins trace back over 2,000 years, with roots in China before spreading to Europe, where it became a staple. The basic ingredients of sauerkraut are simple: finely sliced cabbage and salt. The cabbage is mixed with salt and then it is tightly packed into a fermentation vessel, such as a ceramic crock or glass jar. Water is extracted from the cabbage by the salt, creating a brine that submerges the cabbage, creating an anaerobic environment conducive to fermentation. Over the course of several weeks, naturally occurring lactic acid bacteria on the cabbage leaves ferment the sugars present in the cabbage into lactic acid, preserving the cabbage and giving sauerkraut its distinctive sour flavor.

The preparation of sauerkraut varies from region to region, with some recipes including additional ingredients such as caraway seeds, apples, or juniper berries for flavor variation. Regardless of the additions, the fundamental process remains a remarkable demonstration of controlled microbial activity, resulting in a product that is far greater than the sum of its parts. Sauerkraut is celebrated not only for its tangy taste but also for its health benefits, including enhanced digestion due to its high fiber content and probiotics, vitamins C and K, and antioxidants.

On the other hand, Kimchi is a cornerstone of Korean cuisine, with a history that dates back to the first century BCE. This spicy, fermented vegetable dish encompasses various ingredients, with Napa cabbage and Korean radishes being the most common bases. The preparation of kimchi involves salting the vegetables to remove excess water, then seasoning them with a mix of gochugaru (Korean chili powder), garlic, ginger, scallions, and often fish sauce or fermented seafood for umami depth. This mixture is then tightly packed into containers to ferment at room temperature for several days before being stored in cooler conditions to slow the fermentation process.

The diversity of kimchi is vast, with hundreds of variations that reflect regional preferences, seasonal ingredients, and family traditions. Each batch of kimchi is unique, yet all share the commonality of promoting gut health through a rich supply of probiotics. Kimchi also consists of vitamins A, B, and C, and its ingredients have been linked to health benefits such as reduced cholesterol and antioxidant properties.

The cultural significance of sauerkraut and kimchi extends beyond their health benefits and into the realm of national identity and heritage. In Germany and Eastern Europe, sauerkraut is more than just a food item; it is a link to the past, a symbol of survival and resilience. It is featured prominently in many traditional dishes, celebrated in festivals, and has been carried by explorers and soldiers as a source of nutrition and comfort.

Similarly, kimchi is deeply ingrained in Korean culture, representing an essential element of Korean identity. The making of kimchi, known as "kimjang," involves communal activities that strengthen family and community bonds. Kimchi is present at nearly every meal, reflecting the Korean philosophy of balance and harmony in food. The UNESCO Intangible Cultural Heritage list recognizes Kimjang, emphasizing its cultural importance.

In modern culinary practices, both sauerkraut and kimchi have transcended their traditional roles, finding their way into fusion cuisines and innovative dishes worldwide.

Their complex flavors and health benefits appeal to contemporary palates, while their traditional preparation methods connect eaters to the cultural and historical contexts from which they emerged.

In conclusion, sauerkraut and kimchi exemplify the transformative power of fermentation, turning simple vegetables into complex, flavorful, and nutritious foods that carry the weight of cultural history and tradition. These fermented vegetables have stood the test of time, evolving with changing dietary trends while remaining anchored in their cultural roots. As global interest in fermentation grows, sauerkraut and kimchi continue to be celebrated for their distinctive tastes, health benefits, and the deep cultural significance they hold. Whether enjoyed as part of traditional dishes or incorporated into modern culinary creations, sauerkraut and kimchi stand as enduring symbols of the art and science of fermentation, uniting people across cultures and generations through the shared experience of food.

# CHAPTER VI

# Advanced Fermentation

## Exploring wild fermentation

Wild fermentation is a fascinating and ancient practice that captures the essence of culinary artistry and microbial science. Unlike controlled fermentation, which relies on introducing specific starter cultures, wild fermentation harnesses the diverse array of naturally occurring microbes present in the environment, on the surfaces of fruits and vegetables, and in the air around us. This method allows for a spontaneous and dynamic fermentation process, resulting in unique flavors, textures, and nutritional profiles. This section explores the world of wild fermentation, its principles, benefits, and the rich diversity it brings to the culinary table.

At the heart of wild fermentation is the principle of biodiversity. The environment is teeming with a complex community of bacteria, yeasts, and fungi, each playing a pivotal role in fermentation. When vegetables, fruits, or grains are left to ferment naturally, they become a canvas for these microbes to interact, compete, and collaborate, transforming the substrate into something new and vibrant. This process is not only a testament to the power of nature but also an expression of the specific terroir, considering that the microbial communities can differ significantly between locations, imbuing the fermented products with a sense of place.

Wild fermentation has deep roots in human history, with evidence of its practice dating back thousands of years across various cultures and civilizations. Wild fermentation has been an integral part of human survival and culinary evolution, from the spontaneously fermented beers of ancient Mesopotamia to the diverse pickled vegetables found in traditional diets worldwide. It has enabled our ancestors to preserve the harvest, enhance the nutritional value of their food, and develop complex flavors that are impossible to replicate with industrial food processing techniques.

One of the most compelling aspects of wild fermentation is its accessibility. It democratizes the art of fermentation, requiring no specialized equipment or starter cultures. All that is needed is fresh produce, salt, and a suitable container. This simplicity, however, belies the complexity of the microbial interactions that unfold during fermentation. Numerous factors, including temperature, humidity, and the intrinsic qualities of the fermentation medium influence the process. Mastering wild fermentation becomes an exercise in observation, experimentation, and patience, as the fermenter learns to create the optimal conditions for beneficial microbes to thrive.

Wild fermentation is not without its challenges. The lack of control over the specific strains of microbes that dominate the fermentation can sometimes lead to unpredictable results, with flavor, texture, and safety variations. To mitigate these risks, practitioners of wild fermentation develop a keen sense of smell and taste, learning to determine the signs of a healthy ferment and identifying potential spoilage. This intimate engagement with the process fosters a deep connection to the food, encouraging a mindful consumption approach often lost in modern food systems.

The benefits of wild fermentation extend beyond the culinary. Fermented foods are well-known for their health benefits, particularly regarding gut health. The diverse probiotics generated through wild fermentation can contribute to a balanced gut microbiome, enhancing digestion, boosting the immune system, and potentially mitigating the risk of various chronic diseases. Moreover, the process of fermentation can increase the bioavailability of the nutrients in the food, making them more accessible to the body.

Wild fermentation also embodies the principles of sustainability and resilience. By relying on natural processes and local microbial ecologies, it minimizes the need for external inputs and reduces food waste. The practice encourages local food systems and biodiversity in terms of microbial life and the variety of fermented products it generates. As such, wild fermentation represents a powerful tool for ecological sustainability and food sovereignty, reconnecting people with the natural world and the cycles of growth and decay.

In exploring the world of wild fermentation, one encounters a rich tapestry of flavors, textures, and traditions. Each jar of sauerkraut, kimchi, or sourdough bread is a testament to the complexity of microbial life and the human ingenuity that has harnessed it for nourishment and pleasure. Wild fermentation invites us to embrace the unpredictable, engage with the living world in a meaningful way, and savor the delicious diversity it offers.

In conclusion, wild fermentation is more than just a method of food preservation; it is a celebration of life, diversity, and cultural heritage. It challenges us to relinquish some degree of control, trust in nature's wisdom, and be open to the surprises it may bring. As we continue to explore and understand the microbial world, wild fermentation is a reminder of our interconnectedness with the environment as well as the profound impact of invisible forces on our lives and well-being. Whether as a hobbyist, a culinary professional, or simply an adventurous eater, delving into the practice of wild fermentation opens up a world of possibilities, inviting us to explore, experiment, and enjoy the boundless creativity of nature.

## Using starter cultures and fermentation cultures

In fermentation, the use of starter and fermentation cultures plays a pivotal role in shaping the flavor, texture, and safety of fermented foods. These cultures, consisting of beneficial bacteria, yeasts, or molds, initiate and guide the fermentation process, transforming raw ingredients into complex, flavorful, and often more nutritious foods. This section delves into the intricacies of using starter and fermentation cultures, exploring their significance, varieties, and application in various fermentation practices.

At its core, fermentation is known as a metabolic process where in microorganisms convert carbohydrates, such as sugars and starches, into alcohol or acids. This process will not only preserves the food but it also enhances its nutritional value, digestibility, and flavor profile. Starter cultures and fermentation cultures are integral to this process, providing the necessary microorganisms to ensure a consistent and controlled fermentation.

Starter cultures are specific strains of bacteria, yeast, or mold, carefully selected for their ability to ferment food in a desired manner. They are introduced to the food to kickstart fermentation, ensuring it proceeds efficiently and predictably. The use of starter cultures is widespread in producing yogurt, cheese, sourdough bread, and alcoholic beverages, where consistency in flavor, texture, and safety is paramount.

Fermentation cultures, on the other hand, may refer to the broader community of naturally occurring microorganisms that contribute to the fermentation process. In wild fermentation, for example, vegetables are left to ferment solely with the bacteria present on their surface and in the environment. This can result in a more diverse array of flavors and textures, as the specific mix of microorganisms can vary greatly depending on factors such as the ingredients' origin, the environment, and the season.

One of the main advantages of using starter cultures is the level of control they offer. By selecting a specific culture with known characteristics, fermenters can achieve consistent results, replicate traditional flavors, and reduce the risk of spoilage or the development of harmful bacteria. This is particularly important in commercial fermentation, where product consistency is crucial for consumer satisfaction and brand reliability.

Furthermore, using starter cultures can enhance the health benefits of fermented foods. Many starter cultures are selected for their probiotic properties, contributing to gut health and overall well-being. Introducing these beneficial bacteria can make fermented foods a source of nutrition and a functional food with specific health benefits.

The selection of starter cultures depends on the fermented food type and the desired end product. For instance, Lactobacillus bulgaricus and Streptococcus thermophilus are commonly used in yogurt production for their ability to produce lactic acid and characteristic yogurt flavor. In cheese making, different cultures are used to create the vast array of cheese types, from the soft and creamy Brie to the hard and tangy Cheddar. Each culture contributes its unique flavor and textural properties to the cheese.

The use of fermentation cultures, particularly in wild fermentation, explores the diversity of flavors and textures that can be achieved through fermentation. It represents a more traditional approach, relying on the natural environment and the skill of the fermenter to guide the process. While this method may produce less consistent results, it allows for a greater expression of terroir—the unique characteristics imparted by the local environment—and fosters a deeper connection to the food and its origins.

Despite the benefits, the use of starter cultures and fermentation cultures requires careful consideration and management. Factors such as temperature, pH, and salinity must be controlled to create an environment conducive to the desired microorganisms. Additionally, the quality and purity of the cultures are paramount to prevent contamination and ensure the safety of the fermented product.

In conclusion, the use of starter cultures and fermentation cultures is a fundamental aspect of fermentation, offering both the means to preserve food and enhance its flavor, texture, and nutritional value. Whether opting for the consistency and control provided by starter cultures or embracing the diversity and tradition of wild fermentation, understanding and harnessing these cultures is key to successful fermentation. As interest in fermented foods continues to grow, so too does the appreciation for the art and science of fermentation, a testament to the enduring human fascination with transforming the ordinary into the extraordinary through the power of microorganisms.

## Fermented beverages (kombucha, kefir)

With their rich history and diverse cultural backgrounds, fermented beverages have become increasingly popular in the health and wellness community for their probiotic benefits and unique flavors. Kombucha and kefir, in particular, are two of the most recognized and consumed fermented drinks worldwide. These beverages offer a refreshing taste and bring myriad health benefits, making them a favored choice for those looking to incorporate fermented foods into their diet. This section explores the fascinating world of kombucha and kefir, delving into their origins, production processes, health benefits, and the nuances that make each drink distinct.

The fermented tea beverage known as kombucha, sometimes referred to as the "tea of immortality," has been a part of Chinese culture for millennia. It is produced by fermenting sweetened tea with a symbiotic culture of bacteria and yeast (SCOBY), which creates a film that resembles mushrooms on the liquid's surface. This process typically takes about 7 to 14 days and occurs in two stages. In the first stage, the SCOBY metabolizes the sugar in the tea, producing ethanol and acetic acid, which give kombucha its characteristic sour taste. The second stage involves optional flavoring, where fruits, herbs, or spices can be added to the fermented tea for a secondary fermentation, enhancing its taste and carbonation.

Kefir, on the other hand, originates from the North Caucasus mountains and has been part of the diet of people in this region for centuries. It is produced by fermenting milk (cow, goat, or sheep) with kefir grains, which are not grains in the traditional sense but rather a complex matrix of bacteria and yeasts living in a symbiotic relationship. The fermentation process for kefir is quicker than that of kombucha, usually taking around 24 hours. During this time, the microorganisms in the kefir grains metabolize the lactose in the milk, producing lactic acid, alcohol, and carbon dioxide, resulting in a tangy, slightly effervescent beverage. Kefir can also be produced with non-dairy alternatives, such as coconut water or fruit juices, making it accessible to those with dairy intolerances.

The health benefits of kombucha and kefir are extensive and have been the subject of numerous scientific studies. Both beverages are rich in probiotics, the beneficial bacteria that play a vital part in digestive health, improving digestion and balancing the gut microbiome. These probiotics can also enhance the immune system, reduce inflammation, and may even positively affect mental health through the gut-brain axis. In addition to probiotics, kombucha contains antioxidants from the tea, which can help detoxify the body and protect against disease. As a fermented milk product, Kefir is an excellent source of nutrients, including protein, calcium, B vitamins, and potassium.

Despite their health benefits, consuming these fermented beverages in moderation is important. Kombucha, for example, contains trace amounts of alcohol as a byproduct of fermentation, though it is usually less than 0.5% alcohol by volume when properly prepared. Some commercial brands may have higher alcohol levels, so it is essential to check labels. Kefir, mainly when made from dairy, can be high in calories as well as fat, depending on the milk used, so those watching their calorie intake should consider this.

The production and consumption of kombucha and kefir also reflect a growing interest in DIY fermentation and the movement towards more natural, probiotic-rich foods. Many enthusiasts brew kombucha or ferment kefir at home, experimenting with different teas, sugars, milks, and flavorings to create customized beverages that cater to their taste preferences and health needs. This hands-on approach allows individuals to control the ingredients and fermentation process and connects them to the ancient traditions of food preservation and preparation.

In conclusion, kombucha and kefir represent two pillars of the fermented beverage world, each with its unique history, fermentation process, and health benefits. Their popularity is a testament to the increasing awareness of the importance of probiotics and fermented foods in a balanced diet. As more people seek natural, health-promoting foods, the interest in kombucha, kefir, and other fermented beverages will likely continue growing. Whether commercially produced or home-brewed, these drinks offer a delicious and healthful addition to the modern diet, bridging the gap between culinary tradition and contemporary wellness trends.

## Troubleshooting common fermentation problems

Fermentation, an age-old culinary practice, is crucial in enhancing various foods' nutritional value, flavor, and shelf life. While the process can often be straightforward, specific common issues may arise, deterring both novices and experienced fermenters alike. Understanding how to troubleshoot these problems is essential for guaranteeing the success and safety of the fermentation process. This section delves into some of the most common fermentation problems, offering insights and solutions to help fermenters overcome these hurdles.

One frequent issue encountered during fermentation is the development of mold. Mold can appear on the surface of ferments, especially in open-air fermentations like sauerkraut or kimchi, when the anaerobic (oxygen-free) environment is compromised. This often results from the fermenting food being exposed to air because it is not fully submerged in brine. To prevent mold growth, it's crucial to ensure that the vegetables are completely submerged under the brine using weights. If mold does appear, removing the affected area promptly and ensuring the rest of the ferment is still submerged can salvage the batch, but any signs of mold beneath the surface mean the entire batch should be discarded.

Another common problem is the presence of Kahm yeast, a thin, white film that can form on the surface of fermentations, such as pickles or fermented hot sauces. While not harmful, Kahm yeast can impart an off-flavor to the ferment. Maintaining a proper seal on the fermentation vessel to limit oxygen exposure and ensuring cleanliness can help prevent its occurrence. If Kahm yeast does develop, skimming it off the surface and ensuring the ferment is fully submerged may help, but if the flavor is significantly affected, it might be best to start over.

Unpleasant odors can also be a concern, particularly for those new to fermentation. While fermenting foods, especially vegetables, can produce strong smells due to the release of sulfur compounds, these are generally normal and not indicative of spoilage. However, a putrid or rotten smell is a red flag, signaling that the fermentation has gone awry, possibly due to contamination by unwanted bacteria. In such cases, discarding the batch is advisable to avoid health risks.

Another issue that can arise during fermentation is insufficient acidity, leading to a failed fermentation where the environment is not acidic enough to preserve the food safely. This can be due to several factors, including incorrect salt concentration, temperatures that are too cold for the bacteria to work effectively, or too little starter culture. Addressing these issues might involve adjusting the salt concentration, ensuring the ferment is kept at a suitable temperature, or using a reliable starter culture to kickstart the fermentation process.

Fermenters might also encounter overly soft or mushy textures in vegetables, which is often undesirable. This can result from enzymatic breakdown or over-fermentation. To prevent this, adding tannin-rich leaves like grape, oak, or horseradish to the ferment can help maintain crunchiness. Additionally, closely monitoring the fermentation and refrigerating the product once the desired acidity and texture are achieved can help preserve the ferment's quality.

Another common issue is a lack of fizz in fermented beverages like kombucha or kefir, typically indicative of insufficient fermentation time or a lack of fermentable sugars for the yeast to take in. Ensuring the fermentation is allowed enough time and adjusting the sugar content if necessary can help achieve the desired carbonation.

Additionally, ensuring airtight conditions during the secondary fermentation phase can promote carbonation. Fermentation that progresses too slowly or stalls can be frustrating, often due to temperatures that are too low, an imbalanced ratio of salt to vegetables, or inadequate starter culture. Addressing the specific cause—by moving the ferment to a warmer spot, adjusting the salt concentration, or adding more starter culture—can often revive a sluggish fermentation.

Conversely, too rapid fermentation can lead to over-fermentation, where flavors become overly sour or textures degrade. This is often a result of too warm fermentation conditions. Moving the fermentation to a cooler location can help slow the process, allowing for better control over the final product's flavor and texture.

In conclusion, while fermentation is generally a robust and forgiving process, encountering problems is not uncommon. Understanding these common issues and troubleshooting them can significantly enhance the fermentation experience, ensuring the production of delicious, safe, and nutritious fermented foods and beverages. Patience, observation, as well as a willingness to learn from each batch are key to mastering the art of fermentation. With these troubleshooting tips, fermenters can navigate the challenges of fermentation, enjoying the myriad benefits this ancient culinary practice has to offer.

# CHAPTER VII

# Health Benefits of Pickling and Fermentation

## Nutritional advantages of pickled and fermented foods

The world of pickled and fermented foods is a treasure trove of flavors and a significant source of nutritional benefits. These traditional methods of food preservation, practiced globally for centuries, transform fresh ingredients into products that offer enhanced health benefits. Pickling, through the use of vinegar or brine, and fermentation, utilizing the action of beneficial bacteria, yeasts, and molds, create foods that are rich in vitamins, minerals, probiotics, and enzymes. This section delves into the nutritional advantages of pickled and fermented foods, shedding light on how these processes contribute to a healthier diet and improved well-being.

At the heart of fermented foods is the presence of live microorganisms known as probiotics. These beneficial bacteria are crucial for gut health, playing a pivotal role in digesting food, absorbing nutrients, and bolstering the immune system. Fermented foods like yogurt, kefir, sauerkraut, and kimchi introduce a diverse array of probiotics into the diet, promoting a balanced gut microbiota. This balance is essential for maintaining digestive health, preventing the overgrowth of harmful bacteria, and may even contribute to a reduced risk of certain gastrointestinal conditions which includes irritable bowel syndrome (IBS) and inflammatory bowel disease (IBD).

Moreover, the fermentation process can enhance the nutritional profile of foods. During fermentation, beneficial bacteria produce enzymes and bioactive compounds, including B vitamins (notably B12 in certain fermented soy products), vitamin K2, and omega-3 fatty acids, which are important for different bodily functions, including brain health, blood clotting, and reducing inflammation. For example, the fermentation of soybeans to produce tempeh or natto increases the bioavailability of isoflavones, compounds associated with reduced risks of heart disease and certain cancers.

Fermented foods are also renowned for their enzyme content. These naturally occurring enzymes not only aid in the breakdown of nutrients, making them more digestible but also alleviate the body's digestive workload. This can be especially beneficial for people who have digestive problems, as it can help absorb nutrients from food more efficiently. The improved digestibility of fermented foods is a boon for overall nutrition, allowing the body to assimilate more of the vitamins and minerals present in the food.

Pickling, while often involving the addition of vinegar, can also include fermentation as a method of acidification. This process, known as lacto-fermentation, results in foods that share many of fermented foods' probiotic and enzymatic benefits. Even in vinegar-based pickles, the high content of acetic acid has been shown to have health benefits, including enhanced blood sugar control, weight management, and cardiovascular health. Furthermore, pickled vegetables retain most of their vitamins and minerals, and the acidic environment can sometimes enhance the bioavailability of these nutrients.

Both pickled and fermented foods add dietary diversity in terms of both flavor and nutrition. The addition of spices and herbs to these foods not only contributes to their distinctive tastes but also introduces additional antioxidants and nutrients. For instance, turmeric, commonly added to pickled products, offers anti-inflammatory properties, while the garlic often found in kimchi is known for its immune-boosting effects.

It is important to note that while pickled and fermented foods offer numerous health benefits, moderation is key. Some pickled foods can be high in sodium, which may contribute to increased blood pressure in some individuals. Similarly, certain fermented foods may contain high levels of histamine, which can affect those with histamine intolerance. Therefore, incorporating these foods into a balanced and varied diet is the optimal approach to maximize their health benefits.

In conclusion, the nutritional advantages of pickled and fermented foods are manifold, offering an array of health benefits that complement a balanced diet. From the probiotics that support gut health and immunity to the enzymes that enhance digestion, and the vitamins and minerals that are vital for bodily functions, these foods are nutritional powerhouses. Their role in traditional diets underscores the wisdom of ancient food preservation methods, extending the shelf life of produce and enriching our diet with essential nutrients. As interest in health and nutrition grows, pickled and fermented foods stand out for their contributions to a healthful and flavorful diet, embodying a perfect blend of taste and nutrition.

## Gut health and probiotics

The exploration of gut health and the role of probiotics found in pickled and fermented foods is a subject of increasing interest within both the scientific community and among health-conscious individuals. This fascination is not without reason; the digestive system is intricately linked to various aspects of health, from the immune system to mental well-being. Pickled and fermented foods, revered for their distinct flavors and preservation benefits, are also rich sources of probiotics, the beneficial bacteria that is crucial in maintaining gut health. This section delves into the significance of these microorganisms, their benefits to the digestive system, and how incorporating pickled and fermented foods into the diet can take part to overall health and well-being.

Gut health is known as the function and balance of bacteria in the numerous parts of the gastrointestinal tract. Ideally, organs including the esophagus, stomach, and intestines work together to permit us to eat and digest food without discomfort. But the gut is not only about digestion. It also contains an essential component of the immune system. The gut flora, or gut microbiota, comprises trillions of microorganisms, including bacteria, viruses, fungi, and other microscopic living things. These microorganisms are essential for digestion, help absorb nutrients, and synthesize specific vitamins. The balance of these microorganisms is crucial; an imbalance may lead to digestive disorders, immune dysfunction, and increased susceptibility to infections.

Probiotics are live microorganisms that, when given to the host in sufficient quantities, cause the host to experience a positive impact on their health. They are often called "good" or "beneficial" bacteria because they help keep the gut healthy. Probiotics can regenerate the natural balance of gut bacteria, leading to improved digestion, enhanced immune function, and a reduced risk of many diseases. Fermented foods are natural sources of probiotics due to the action of lactic acid bacteria, among the most common microbes in the fermentation process.

Fermented foods like sauerkraut, kimchi, yogurt, kefir, and kombucha have been part of human diets for centuries, primarily for their preservation properties and unique tastes. More recently, the scientific community has started to comprehend the role that they play in fostering gut health. The fermentation process involves the conversion of both starches and sugars into lactic acid by lactic acid bacteria. This process results in the creation of an acidic environment that impedes the growth of bacteria that are harmful to the organism. After going through this process, the food is preserved and then enriched with live cultures of probiotics.

Ingestion of fermented foods has the potential to increase the diversity and density of the microbiota found in the gut. There is a correlation between having a diverse gut microbiota and having better health. This includes a reduced risk of gastrointestinal conditions such as IBS (irritable bowel syndrome) and IBD (inflammatory bowel disease), as well as a lower risk of systemic conditions like obesity and type 2 diabetes. Probiotics from fermented foods can also strengthen the intestinal barrier function, reducing the likelihood of pathogens entering the bloodstream and causing infection.

Moreover, fermentation can increase nutrient bioavailability, making fermented foods an even more valuable part of the diet. For example, fermentation can increase levels of specific B vitamins and omega-3 fatty acids, essential nutrients that support overall health. Additionally, some fermented foods contain dietary fibers and prebiotics, which nourish the helpful bacteria in the gut, further promoting a healthy digestive system.

Despite the clear benefits, it's essential to approach the consumption of pickled and fermented foods with balance and awareness. Not all pickled foods are fermented, and not all fermented foods contain live probiotics. For instance, pickles made with vinegar rather than through natural fermentation do not contain beneficial bacteria. Similarly, some commercially fermented foods are pasteurized, a process that kills bacteria, including any probiotics. Therefore, for gut health benefits, it's crucial to choose naturally fermented products that specify they contain live and active cultures.

In conclusion, the probiotics found in pickled and fermented foods significantly promote gut health, offering a wide array of benefits from improved digestion and enhanced immune function to a potential reduction in chronic disease risk. Incorporating an array of these foods into the diet can be a delicious as well as effective way to support a healthy, balanced gut microbiota. As research into the human microbiome continues to evolve, the traditional wisdom of consuming fermented foods for health finds new grounding in scientific evidence, underscoring the interconnectedness of diet, gut health, and overall well-being.

## Immune system support

The interconnection between diet and health has long been recognized, with an increasing focus on the specific role of pickled and fermented foods in supporting the immune system. These foods, which undergo a process that encourages the growth of beneficial microbes, are a means to preserve perishables and a potential key to enhancing bodily defenses against pathogens. This section explores how pickled and fermented foods may bolster the immune system, drawing upon scientific research and traditional knowledge to illuminate the significance of these dietary components.

The relationship between pickled and fermented foods and immune health lies in the concept of the gut microbiota, an intricate community of microorganisms residing in the digestive tract. This microbiota plays several critical roles in health and disease, including the metabolism of nutrients, protection against pathogens, and modulation of the immune system. The consumption of fermented foods, which are rich in probiotics, takes part to the diversity and balance of the gut microbiota, which in turn influences the immune system's efficacy.

Fermented foods like yogurt, sauerkraut, kefir, kimchi, and miso are sources of live bacteria that can colonize the gut, contributing to its microbial diversity. These beneficial bacteria compete with potential pathogens for resources and space, reducing the likelihood of pathogenic overgrowth. Furthermore, they can enhance the integrity of the gut barrier, preventing harmful substances from leaking into the bloodstream and triggering immune responses. This barrier function is crucial for maintaining immune health, as it shields the body from external invaders and reduces inflammation.

Beyond reinforcing the gut barrier, the probiotics found in fermented foods can directly modulate the immune system. They have been shown to influence the activity of various immune cells, enhancing the body's ability to fend off infections. For example, specific Lactobacillus and Bifidobacterium strains, commonly found in fermented dairy products, can stimulate the production of antibodies and activate macrophages and natural killer cells, key players in the immune response. These interactions suggest that regular consumption of fermented foods could enhance the body's resilience to infections. Although not always fermented, pickled foods can also contribute to immune support, mainly when they are produced through fermentation rather than vinegar-based pickling. The lactic acid bacteria included in the fermentation of pickles produce metabolites with antimicrobial properties, further supporting the body's defense mechanisms. Additionally, the vitamins and antioxidants preserved or even enhanced through the pickling process, such as vitamin C and beta-carotene, are vital for the proper functioning of the immune system.

The impact of fermented foods on the immune system extends to the production of short-chain fatty acids

(SCFAs) by the fermentation of dietary fibers by gut bacteria. SCFAs, including butyrate, propionate, and acetate, have been shown to have anti-inflammatory effects and to play a role in regulating the immune response. By influencing the production of these compounds, consuming fermented foods may help modulate inflammation and immune regulation, offering protection against autoimmune and inflammatory diseases.

It is important to note that the benefits of pickled and fermented foods on immune health can change depending on the type of food, the strains of bacteria involved, and the individual's gut microbiota composition. Not all fermented foods contain live probiotics; pasteurization can destroy these beneficial microbes. Therefore, for maximum health benefits, it is advisable to choose unpasteurized fermented foods that are known to contain live and active cultures, and to incorporate a variety of such foods into the diet to capture a broad range of bacteria strains.

In addition to their direct effects on the gut microbiota and immune cells, pickled and fermented foods can also influence overall health and immune function through their nutritional content. Many of these foods are rich in essential nutrients, including vitamins B and K, minerals such as iron and magnesium, and dietary fiber, all of which are important for maintaining a healthy immune system.

In conclusion, the consumption of pickled and fermented foods offers several potential benefits for immune system support. By maintaining a healthy gut microbiota, direct modulation of immune cell activity, and provision of essential nutrients, these foods can contribute to a stronger, more responsive immune system. As research in this area continues to evolve, understanding how best to harness the immune-supportive properties of pickled and fermented foods will likely grow, reinforcing the importance of these traditional dietary practices in modern health and nutrition. Incorporating various fermented and pickled foods into the diet represents a simple, yet potentially impactful, strategy for enhancing immune health and overall well-being.

## Weight management benefits

The global interest in health and wellness has spotlighted the role of diet in managing weight, with pickled and fermented foods emerging as potential allies in this ongoing battle against obesity. These traditional food preservation methods not only extend the foods' shelf life and enhance their flavors but also offer a variety of health benefits, including support for weight management. This section explores the mechanisms through which pickled and fermented foods may contribute to weight loss and maintenance, underscoring the importance of these foods in a balanced diet.

Fermented foods, including yogurt, kefir, kimchi, and sauerkraut, undergo a metabolic process where microorganisms, primarily bacteria and yeasts, convert sugars and starches into alcohol or organic acids. This process not only preserves these foods but also enriches them with probiotics, live microorganisms that offer health advantages when consumed in adequate amounts. The impact of these probiotics on weight management has been the subject of many studies, revealing several mechanisms through which they may exert their effects.

One of the primary ways fermented foods can aid in weight management is through improving gut health. The gut microbiota is crucial in metabolism and weight regulation, with imbalances associated with obesity and other metabolic disorders. Fermented foods contribute to a more diverse and balanced gut microbiota, which can enhance the body's metabolism and energy expenditure. Probiotics from fermented foods can also influence the absorption of dietary fats, increasing the amount excreted rather than stored in the body. A healthier gut microbiota can also improve insulin sensitivity, reducing the risk of obesity-related conditions such as type 2 diabetes.

Another mechanism is the potential appetite-regulating effect of fermented foods. Some studies suggest that consuming probiotic-rich foods can increase the production of appetite-regulating hormones such as GLP-1 (glucagon-like peptide-1) and PYY (peptide YY), leading to lowered calorie intake and, ultimately, weight loss. These foods may also help reduce levels of the hunger hormone ghrelin, further curbing appetite and aiding in weight control.

Furthermore, fermentation can increase certain nutrients' bioavailability, making fermented foods particularly nutrient-dense. Incorporating fermented foods into the diet can help ensure that nutritional needs are met without the need for excessive calorie intake, supporting weight management efforts. For instance, fermentation can increase levels of B vitamins, essential for energy metabolism, and improve the bioavailability of minerals like iron and zinc, which are crucial for maintaining bodily functions that support a healthy weight.

While not always fermented, pickled foods can also play a role in weight management, primarily due to their low calorie content and the potential for increasing satiety. Foods like pickled cucumbers, beets, and carrots are high in water and fiber but low in calories, making them excellent choices for snacking or adding bulk to meals without significantly increasing calorie intake. The tangy flavor of pickled foods can also enhance the palatability of dishes, making it easier to enjoy a variety of nutrient-dense, low-calorie foods as part of a weight management diet.

However, it is essential to note that not all pickled and fermented foods are created equal in terms of their health benefits. Some commercially prepared pickled foods are high in sodium and added sugars, which can negate their potential benefits for weight management. Similarly, some fermented foods may be high in calories and fats, depending on how they are prepared. Therefore, it is crucial to choose natural, low-sugar, and low-sodium options and to be mindful of portion sizes when incorporating these foods into the diet.

In conclusion, pickled and fermented foods offer several benefits for weight management, from improving gut health and metabolism to regulating appetite and increasing nutrient density. Their inclusion in a balanced diet can support efforts to maintain a healthy weight, alongside other lifestyle factors including regular physical activity and adequate sleep. As with any dietary approach, moderation and variety are key; incorporating a wide range of fermented and pickled foods can help maximize their health benefits while ensuring a balanced intake of nutrients. By understanding and leveraging the weight management benefits of these traditional foods, individuals can enjoy delicious, flavorful options that support their health and wellness goals.

# CHAPTER VIII

# Pickling and Fermentation for Long-Term Storage

## Strategies for preserving food during emergencies

In the face of emergencies, whether due to natural disasters, economic downturns, or global pandemics, the importance of food preservation becomes paramount. The ability to keep food for longer periods can be a lifeline, ensuring nutritional needs are met even when access to fresh supplies is compromised. This section explores various strategies for preserving food during emergencies, focusing on methods that are practical, effective, and accessible to most households.

Canning is a widely embraced method for preserving various foods, including fruits, vegetables, meats, and soups. This technique involves placing food in jars and/or cans and then heating them to a temperature that destroys microorganisms and also inactivates enzymes that could lead to spoilage. Once sealed, canned goods can last for years, retaining much of their nutritional value. While the process requires some initial investment in equipment and a learning curve to master safe canning practices, the ability to create a long-term stockpile of diverse food items makes it invaluable during emergencies.

Drying, or dehydrating food, is one of the oldest and simplest preservation methods. By removing moisture, the growth of microorganisms is inhibited, significantly extending the shelf life of foods. Fruits, vegetables, meats, and herbs can all be dried and stored for months or even years. Modern food dehydrators make the process more efficient, but drying can also be accomplished using an oven or air-drying in a warm, dry environment. Dried foods are lightweight and space-saving, making them ideal for emergency kits and limited storage spaces.

Freezing is another straightforward method for preserving food, with the cold temperatures slowing down the activity of spoilage-causing organisms and enzymes. Almost any food can be frozen, including bread, dairy products, cooked meals, meats, and produce. The key to successful freezing is proper packaging to prevent freezer burn and maintain a consistent temperature. While freezing offers a convenient way to store perishable items, it depends on a continuous power supply, which may be a consideration during prolonged emergencies.

Fermentation is a natural method that preserves food and enhances its nutritional content and digestibility. Through fermentation, foods like cabbage, cucumbers, and milk are transformed into sauerkraut, pickles, and yogurt, respectively. These fermented foods are high in probiotics, which support gut health. Fermentation requires minimal equipment and can be done in any kitchen, making it an accessible option for prolonging the storage life of fresh produce and dairy.

Similar to fermentation, pickling involves immersing foods in an acidic solution, usually vinegar, sometimes with added salt, sugar, and spices. This acidic environment impedes the growth of spoilage-causing bacteria. Pickled vegetables, fruits, and even eggs can provide a welcome variety in diet during emergencies. Unlike fermentation, pickling does not necessarily produce probiotics, but it still offers a means to safely preserve a wide range of foods.

Root cellaring is a traditional method for storing root vegetables, such as potatoes, carrots, onions, beets, and hardy fruits like apples and pears. A root cellar can be as straightforward as a buried container or a specially constructed underground room that maintains a cool, stable temperature and high humidity. This method leverages the natural keeping qualities of these foods, allowing them to last for several months without the need for electricity.

Vacuum sealing, while a more modern technique, effectively extends the storage life of food by removing air from the packaging. This method reduces oxidation and the growth of bacteria and molds. Vacuum-sealed foods can be stored at room temperature, in the refrigerator, or frozen, further enhancing their preservation. The first investment in a vacuum sealer can be offset by the savings in food costs and the security of having a diverse food supply on hand.

Each of these strategies for preserving food during emergencies has its advantages and limitations. The choice of methods will depend on various factors, including the types of food available, storage space, equipment, and energy sources. Moreover, understanding the nutritional needs of one's household and diversifying preservation techniques can ensure a balanced diet is maintained even in challenging circumstances.

In conclusion, preserving food is a prudent and essential practice for preparing for emergencies. Canning, drying, freezing, fermenting, pickling, root cellaring, and vacuum sealing are all viable strategies for ensuring food security. By mastering these methods, individuals can build resilience against uncertainty and sustain themselves and their loved ones through periods of scarcity.

## Proper storage techniques for pickled and fermented foods

The resurgence of interest in traditional food preservation techniques, particularly pickling and fermentation, has brought with it a need for understanding the proper storage techniques to ensure these foods maintain their quality, safety, and nutritional benefits. Pickled and fermented foods, celebrated for their enhanced flavors, probiotic benefits, and extended shelf life, require specific conditions for storage to preserve these attributes. This section delves into the nuanced world of storing these valuable food items, offering insights into the practices that can maximize their longevity and quality.

Pickling, a method that uses vinegar or a saltwater brine to preserve food, and fermentation, which relies on natural bacteria to create an acidic environment, both inhibit the growth of spoilage-causing bacteria. However, the stability of these preserved foods can be compromised if they are not stored correctly. Understanding the factors that influence their storage life and quality is essential for anyone looking to incorporate these healthful and flavorful foods into their diet.

The environment's temperature is the primary consideration in storing pickled and fermented foods. Most fermented foods thrive and continue to mature in cool conditions, which slow down the fermentation process and help maintain the desired taste and texture. A refrigerator, typically set at or below 4°C (39°F), offers an ideal environment for storing these foods once the initial fermentation phase is complete. This temperature slows down the activity of the bacteria, effectively pausing the fermentation process and preserving the food in its desired state for consumption.

However, not all fermented foods require refrigeration. Some, like sauerkraut and kimchi, can be stored in a cool cellar or pantry at temperatures just above freezing, provided they are kept in airtight containers. This method is beneficial in situations where refrigerator space is limited or for those seeking to adhere to more traditional storage methods. It's crucial, however, to monitor these foods for signs of continued fermentation, such as bubbling or overflow, which may indicate that they should be consumed soon or moved to a cooler environment.

Light exposure is another critical factor to consider. Both pickled and fermented foods are best stored in dark conditions, as light can lead to the degradation of vitamins and can cause certain types of bacteria to proliferate, potentially spoiling the food. Amber glass jars, ceramic containers, or simply storing foods in a dark pantry or cupboard can protect them from light exposure. For those using clear glass containers, wrapping the jars in a cloth or storing them in boxes to shield them from light may be beneficial.

The container's material used for storage also plays a significant part in keeping the quality of pickled and fermented foods. Glass is often the preferred choice for many, as it is non-reactive and does not absorb flavors or odors. It also offers an airtight seal that prevents oxygen from entering and spoiling the food. While plastic containers can be used, they should be food-grade and BPA-free to avoid chemical leaching. Metal containers are commonly not recommended for long-term storage of acidic foods like pickles and fermented vegetables, as the acid can react with the metal, leading to corrosion and contamination of the food.

Proper sealing is essential to prevent contamination and oxidation. For fermented foods, ensuring that the food remains submerged in its liquid (brine or juice) is key to preventing mold and unwanted bacteria from developing. Using weights or a smaller jar inside the fermentation vessel can help submerge vegetables. For pickled foods, ensuring that the jars are sealed tightly after opening and using clean utensils each time to remove the food can prevent contamination and prolong shelf life.

Finally, the duration of storage is an important consideration. While pickling and fermentation extend the shelf life of foods, they are not indefinite solutions. Most pickled and fermented foods are best consumed within a few months to a year of preparation, depending on the specific food and storage conditions. Over time, flavors can change, and nutritional content may diminish. Regularly inspecting stored foods for signs of spoilage, such as off-odors, discoloration, or mold, is crucial in ensuring they remain safe to eat.

In summary, properly storing pickled and fermented foods involves careful consideration of temperature, light exposure, container material, sealing methods, and storage duration. By complying to these guidelines, individuals can enjoy the distinctive flavors, nutritional benefits, and probiotic qualities of these foods long after they have been prepared. As interest in these traditional preservation techniques continues to grow, understanding the nuances of storage becomes essential for anyone looking to integrate these healthful, flavorful foods into their culinary repertoire.

## Shelf life and safety considerations

Preserving food through pickling and fermentation is a practice as ancient as civilization. These methods extend the shelf life of perishable items, enrich flavors, and offer health benefits. However, understanding the shelf life and safety considerations for pickled and fermented foods is essential to ensure they remain safe and enjoyable to eat. This section explores the complexities of preserving food through these methods, shedding light on maximizing shelf life while maintaining safety and quality.

Pickled foods, created by immersing produce in an acidic solution, typically vinegar, or fermenting them in saltwater brine, are staples in many diets worldwide. Fermentation, a related but distinct process, relies on natural bacteria and yeasts to convert sugars into acids, gases, or alcohol. Both processes significantly inhibit the growth of spoilage-causing microorganisms, effectively extending the shelf life of foods. Despite these preservative effects, several factors influence the safety and longevity of pickled and fermented products.

The shelf life of pickled and fermented foods is determined by the preservation method, the ingredients used, the storage conditions, and the container's integrity. Properly canned and sealed pickles can last for up to two years when stored in a cool, dark place. Once opened, however, their shelf life decreases to about a month in the refrigerator. Fermented foods, like sauerkraut and kimchi, continue to evolve in flavor and texture over time due to ongoing microbial activity. While refrigeration slows this activity, these foods are generally best consumed within a few months to a year of fermentation, depending on the specific product and personal taste preferences.

Temperature plays a crucial role in these foods' safety and shelf life. Refrigeration at or below 4°C (39°F) is ideal for slowing down fermentation and preventing the growth of undesirable microorganisms. However, not all pickled and fermented foods require refrigeration. Many can be stored at room temperature until opened, after which refrigeration is necessary to ensure safety. It is important to note that storage conditions should be stable, as fluctuations in temperature can compromise the integrity of the food, leading to spoilage or the proliferation of pathogens.

Light and oxygen are two factors that can adversely affect pickled and fermented foods. Exposure to light can degrade vitamins and encourage the growth of certain bacteria, while oxygen can lead to oxidation and spoilage. Storing these foods in dark, airtight containers can mitigate these risks, preserving their quality and extending their shelf life. Glass jars with tight-sealing lids are commonly used for this purpose, offering the added benefit of allowing visual inspection for signs of spoilage, such as mold growth or gas bubbles.

Safety considerations for pickled and fermented foods also include the potential for contamination by harmful microorganisms if proper hygiene as well as food safety practices are not followed. Botulism, which is caused by bacterium Clostridium botulinum, is a rare but serious risk connected with improperly canned or fermented foods. To minimize this risk, it is crucial to follow tested recipes and canning procedures, ensure the acidity of pickled products is sufficient to inhibit bacterial growth, and use appropriate salt concentrations in fermented foods to promote the development of beneficial bacteria while suppressing harmful ones.

Cross-contamination is another concern, especially when opening and consuming these foods. Using clean utensils each time a product is accessed can prevent the introduction of spoilage-causing bacteria. Regularly inspecting stored pickled and fermented foods for signs of spoilage, including off-odors, color changes, or mold presence, is also essential. If in doubt, discarding the product is safer than risking foodborne illness.

In conclusion, pickled and fermented foods offer a delicious and nutritious way to preserve the bounty of the harvest, with the added benefits of probiotics and enhanced flavors. Understanding the factors influencing their shelf life and safety is key to enjoying these foods at their best. By adhering to proper preparation, storage, and hygiene practices, one can minimize the risks associated with these preservation methods, ensuring that pickled and fermented foods remain a safe and cherished part of the diet. As with all food preservation techniques, respect for the process and attention to detail are paramount in achieving the perfect balance of safety, flavor, and nutritional value.

## Incorporating pickled and fermented foods into emergency rations

In the realm of emergency preparedness, ensuring a diverse and nutritious food supply is paramount. Amid the staples of grains, canned goods, and dried proteins, incorporating pickled and fermented foods into emergency rations offers a strategic advantage. These foods not only provide essential nutrients and probiotics but also introduce variety in flavor, which can be a welcome reprieve in stressful situations. This section explores the benefits and considerations of including pickled and fermented foods in emergency supplies, underlining their value in maintaining health and morale during crises.

Pickled and fermented foods, with their long shelf life and lack of need for refrigeration once sealed, fit seamlessly into emergency food planning. The preservation process of pickling, through vinegar or brine, and fermentation, via the action of beneficial bacteria, naturally extends the usability of fruits, vegetables, and even some proteins. These methods not only inhibit the growth of harmful bacteria but also enhance the nutritional profile of the foods, adding essential vitamins and minerals that might otherwise be scarce in emergency diets.

The probiotic qualities of fermented foods like sauerkraut, kimchi, yogurt, and kefir are particularly beneficial. These foods introduce healthy bacteria to the gut, which can boost the immune system—a critical factor during emergencies when healthcare access may be limited. The fermentation process also expands the bioavailability of nutrients, making these foods even more nutritious. For instance, fermentation can increase levels of B vitamins, essential for energy production and stress management, and improve the absorption of minerals such as iron and calcium.

Incorporating these foods into emergency rations also addresses the psychological aspect of surviving in adverse conditions. The unique and varied flavors of pickled and fermented foods can break the monotony of emergency meals, providing culinary variety that can lift spirits and improve morale. The comfort of enjoying a flavorful meal should not be underestimated in its ability to provide psychological relief and a sense of normalcy amid chaos. Moreover, pickled and fermented foods' versatility enhances their suitability for emergencies. They can be consumed directly from the jar, requiring no preparation or cooking, which is invaluable when resources like fuel and clean water are scarce. They can also be used to complement other foods, adding flavor and nutrition to basic staples. For example, pickled vegetables can accompany rice or beans, and fermented dairy products can be mixed with granola or fruit preserves.

However, including pickled and fermented foods in emergency rations requires thoughtful planning. It's essential to consider the dietary preferences and potential allergies of those who may need to rely on these supplies. Additionally, while pickled and fermented foods can last for months or even years when sealed, once opened, they generally need to be consumed within a relatively short period and kept cool, which could be challenging in some emergency scenarios.

When selecting these foods for emergency rations, opt for varieties that are high in nutritional value and low in added sugars and sodium. Many commercially available pickled products are high in salt, which can result in increased thirst and dehydration if clean water is in short supply. Similarly, some fermented foods may contain high levels of added sugars, which can affect blood sugar levels. Choosing natural, low-sodium, and low-sugar options or preparing homemade versions where possible can mitigate these concerns.

Storage is another consideration. Pickled and fermented foods should be stored in a cool, dark place to keep their quality and extend their shelf life. Glass containers, while ideal for preserving the integrity of these foods, may not be practical for all emergency situations due to their weight and fragility. Although lighter and less prone to breaking, plastic containers should be food-grade and BPA-free to ensure safety.

In conclusion, including pickled and fermented foods in emergency rations offers a multifaceted approach to preparedness. These foods provide essential nutrients and probiotics, contribute to dietary variety, and require no preparation, making them ideal for limited resources. However, successfully integrating these foods into emergency supplies demands careful selection and storage considerations to maximize their benefits while minimizing potential drawbacks. By thoughtfully incorporating pickled and fermented foods into emergency planning, individuals can enhance their resilience, ensuring access to nutritious and enjoyable meals even in the face of adversity.

# CHAPTER IX

# Incorporating Pickling and Fermentation into Prepper Lifestyles

## Integrating pickling and fermentation into meal planning

In the landscape of modern culinary practices, integrating pickling and fermentation into meal planning stands out as a noteworthy trend, blending ancient food preservation techniques with contemporary dietary habits. This marriage of old and new offers not just a pathway to enhancing flavors and diversifying diets but also to reaping nutritional benefits and minimizing food waste. This section explores the multifaceted advantages of incorporating pickled and fermented foods into meal planning, offering insights into how these practices can elevate everyday eating experiences.

The essence of pickling and fermentation lies in their ability to preserve perishable foods by creating an acidic environment, either through the addition of vinegar or the natural production of lactic acid by beneficial bacteria. This process not only extends the shelf life of foods but also enriches them with unique flavors and probiotics, which are beneficial for gut health. Integrating these foods into meal planning opens up a world of culinary possibilities, allowing for the creation of meals that are both flavorful and nutritionally enhanced.

One of the primary benefits of incorporating pickled and fermented foods into meal planning is the introduction of probiotics into the diet. Foods such as kimchi, sauerkraut, kefir, and yogurt are rich in live cultures that can bolster the gut microbiome, improving digestion and potentially enhancing immune function. By planning meals that include these fermented items, individuals can effortlessly incorporate these health benefits into their daily diets, promoting overall wellness.

Moreover, pickled and fermented foods add a depth of flavor that can elevate simple dishes to new heights. The tangy acidity of pickled vegetables can cut through the richness of fatty meats, while the complex flavors of fermented condiments like miso and tempeh can serve as the backbone of savory dishes. Including these elements in meal planning encourages culinary creativity and experimentation, allowing home cooks to explore a range of global cuisines and flavor profiles.

Another significant advantage of integrating pickling and fermentation into meal planning is food waste reduction. Seasonal produce can be preserved at the peak of freshness, ensuring a supply of fruits and vegetables year-round. This minimizes waste and contributes to a more sustainable and eco-friendly kitchen. Meal planners can capitalize on this by incorporating preserved produce into their recipes, ensuring that nothing goes to waste and that meals are enriched with the nutritional benefits of a diverse range of produce.

Incorporating pickled and fermented foods into meal planning also allows for greater flexibility and convenience. Preserved foods have a longer shelf life, reducing the frequency of grocery shopping trips and providing a reliable stock of ingredients that can be used to assemble quick, nutritious meals. This can be a game-changer for busy individuals and families, ensuring that healthful, homemade meals are always within reach, even on the busiest days.

The process of integrating pickling and fermentation into meal planning begins with the preservation of seasonal produce or the purchase of high-quality, store-bought pickled and fermented products. From there, these items can be thoughtfully incorporated into weekly meal plans. For example, pickled vegetables can add brightness to sandwiches and salads, fermented dairy products can serve as the base for smoothies and dressings, and fermented grains and legumes can provide a savory umami flavor to soups and stews.

Creativity is key when integrating these foods into meals. Experimenting with different flavor mixtures and textures can lead to delightful discoveries and can inspire a more adventurous approach to cooking. Additionally, understanding the nutritional content and health benefits of various pickled and fermented foods can guide meal planning decisions, ensuring that meals are delicious but also balanced, and healthy.

In conclusion, integrating pickling and fermentation into meal planning offers a wealth of benefits, from enhancing the flavor and nutritional content of meals to promoting gut health and reducing food waste. By embracing these ancient preservation techniques, home cooks can add variety and depth to their culinary repertoire, explore new cuisines, and enjoy the convenience of having a pantry stocked with versatile, flavorful ingredients. As society continues to seek ways to eat more sustainably and healthfully, the practices of pickling and fermentation stand out as valuable tools in the modern meal planner's toolkit, bridging the gap between tradition and innovation in the kitchen.

## Creating resilient food systems

In the modern quest for sustainability and resilience in food systems, traditional practices such as pickling and fermentation emerge as methods of food preservation and as strategies to strengthen our food security and reduce waste. These age-old techniques, which have nourished civilizations across the globe for centuries, offer valuable lessons in creating resilient food systems that can withstand the challenges of climate change, resource scarcity, and global disruptions. This section explores how integrating pickling and fermentation into our food systems can contribute to resilience, sustainability, and health, reflecting on these practices' environmental, economic, and nutritional impacts.

Pickling and fermentation are processes that extend the shelf life of perishable foods through natural preservation methods. Pickling involves submerging foods in an acidic solution, such as vinegar, or a saltwater brine, creating an environment where spoilage-causing bacteria cannot thrive. Conversely, fermentation relies on beneficial microbes to break down sugars and starches in food, producing acids, alcohol, or gases as byproducts, which act as natural preservatives. These methods not only ensure food safety and prolong shelf life but also enhance the nutritional value and flavor of foods, offering a dual benefit of preservation and improvement.

One of the key contributions of pickling and fermentation to resilient food systems is their ability to reduce food waste. In many parts of the world, a significant portion of food spoils before it can be consumed, contributing to food insecurity and environmental degradation. By preserving seasonal surpluses and less-than-perfect produce that might otherwise be discarded, pickling and fermentation can help minimize waste and maximize the utility of available resources. This practice not only extends the availability of fruits and vegetables beyond their natural growing seasons but also supports local economies by allowing farmers and producers to retain the value of their harvests.

Moreover, pickling and fermentation contribute to food sovereignty and reduce dependence on industrial food production and long supply chains. These techniques can be practiced at home or within local communities, empowering individuals and communities to take control of their food sources. This localized approach to food preservation enhances community resilience, ensuring access to nutritious foods even in times of crisis or when global supply chains are disrupted. By fostering a culture of self-reliance and skill-sharing, pickling and fermentation strengthen the social fabric and support the development of sustainable, community-based food systems.

From an environmental perspective, pickling and fermentation are energy-efficient food preservation methods. Unlike freezing or refrigeration, which require continuous energy input, these techniques rely on natural processes that do not depend on electricity. This low-energy approach reduces the carbon footprint associated with food preservation and aligns with broader goals of reducing energy consumption and minimizing climate change. Furthermore, by enabling the preservation of locally sourced and seasonally available foods, these methods can reduce the carbon emissions connected with transporting food over long distances.

Nutritionally, pickled and fermented foods offer significant health benefits, contributing to the resilience of individuals and populations. The fermentation process, in particular, enhances the bioavailability of nutrients, making it simpler for the body to absorb vitamins and minerals. Fermented foods are also high in probiotics, known as beneficial bacteria that support gut health and strengthen the immune system. Integrating these foods into diets can improve nutritional outcomes, reduce the prevalence of chronic diseases, and lessen the burden on healthcare systems, further contributing to societal resilience.

Despite the benefits, integrating pickling and fermentation into modern food systems presents challenges. These include overcoming cultural barriers, addressing food safety concerns, and ensuring access to knowledge and resources for safe and effective preservation practices. Education and community engagement are crucial in addressing these challenges, as is the development of supportive policies and infrastructure that encourage local food production and preservation.

In conclusion, pickling and fermentation represent more than just methods of food preservation; they are integral components of a resilient and sustainable food system. By reducing food waste, supporting local economies, enhancing nutritional value, and fostering self-reliance, these practices contribute to environmental sustainability, economic stability, and public health. As the world faces increasing food security and sustainability challenges, the wisdom embedded in these traditional techniques offers a path forward, guiding efforts to build more resilient food systems for future generations.

## Community-building through food preservation

The act of preserving food through pickling and fermentation transcends mere sustenance, weaving a rich tapestry of community bonds and cultural heritage. These ancient methods, rooted in necessity, have evolved into powerful tools for community-building, fostering connections that span generations and geographies. This section explores how the communal practices of pickling and fermentation ensure food security and cultivate a sense of belonging, shared identity, and resilience within communities.

Historically, food preservation was a communal affair, with harvests prompting gatherings where everyone participated in the preparation of pickles and fermented goods for the winter months. These gatherings were more than just practical; they were social events that strengthened community ties, shared knowledge across generations, and celebrated the land's bounty. In many cultures, recipes were passed down as heirlooms, embedding family and regional identities into the very flavors of the preserved foods.

In contemporary times, the revival of interest in pickling and fermentation is rekindling these communal practices. Workshops, classes, and shared kitchen spaces have become modern arenas for exchanging knowledge and traditions related to food preservation. These gatherings serve as entry points for community members to engage with one another, learn new skills, and deepen their understanding of pickled and fermented foods' cultural and nutritional aspects. The communal aspect of these practices fosters a sense of cooperation and mutual support, critical elements in building resilient communities.

Community gardens and local food projects have also embraced pickling and fermentation as strategies for maximizing the use of produce, reducing waste, and promoting food sovereignty. By involving community members in the growing, harvesting, and preserving processes, these initiatives create inclusive spaces where people from diverse backgrounds can contribute and benefit. The shared experience of nurturing and then preserving food strengthens community bonds, emphasizing the collective effort required to sustain and nourish one another.

Moreover, the practice of pickling and fermentation can act as a cultural bridge, introducing individuals to the culinary traditions of different cultures through the universal language of food. Fermented foods like kimchi, sauerkraut, and kombucha, each rooted in distinct cultural practices, offer opportunities for cultural exchange and understanding. Community events centered around food preservation can facilitate intercultural dialogue, celebrate diversity, and foster a sense of global connectedness among participants.

Pickling and fermentation's environmental and economic benefits also contribute to community resilience. By preserving seasonal produce, communities can reduce reliance on imported foods, support local agriculture, and minimize their ecological footprint. The low-cost and low-tech nature of these preservation methods make them accessible to a wide range of people, including those with limited resources, further democratizing access to healthy, nutritious food.

Community-building through food preservation also extends to online platforms, where enthusiasts share recipes, techniques, and experiences with a global audience. Social media groups, forums, as well as blogs dedicated to pickling and fermentation have created virtual communities that support and inspire both novices and experts alike. These digital spaces complement physical gatherings, expanding the reach of these traditional practices and fostering a sense of belonging among individuals who may never meet in person.

Despite the many benefits, challenges such as food safety concerns, lack of knowledge, and limited resource access can hinder participation in community pickling and fermentation activities. Addressing these problems demands a concerted effort from community leaders, food safety experts, and local organizations to provide education, resources, and support. Equipping community members with the knowledge as well as skills needed to preserve food safely can overcome these barriers, ensuring that the benefits of these practices are accessible to all.

In conclusion, the revival of pickling and fermentation as communal practices offers profound opportunities for community-building. Beyond the tangible outcomes of food preservation, these activities foster social connections, cultural exchange, and environmental sustainability. By drawing on the collective wisdom of past generations and adapting it to contemporary needs, communities can strengthen their bonds, celebrate diversity, and build resilience. In a world increasingly characterized by isolation and division, the communal act of preserving food stands as a testament to the power of shared effort and tradition to bring people together.

## Recipes and meal ideas for preppers

For preppers, the art of pickling and fermentation is not just a hobby but a strategic approach to ensuring a sustainable, nutritious food supply in times of emergency. These ancient preservation methods extend the shelf life of perishable items and enhance their nutritional value and flavor. This section delves into practical recipes and meal ideas that incorporate pickled and fermented foods, offering preppers innovative ways to diversify their diet and maximize their food reserves.

Starting with the basics, a simple sauerkraut recipe can serve as a cornerstone for a prepper's pantry. Made from just cabbage and salt, sauerkraut undergoes a fermentation process that produces beneficial probiotics. To prepare, thinly slice cabbage, toss it with a small amount of salt, and pack it tightly into a clean jar. The salt draws out the water from the cabbage, creating a brine in which it can ferment. Kept at room temperature for a few days to a few weeks, the cabbage will transform into sauerkraut. Once fermented, it can be stored in a cool, dark place for several months. This tangy condiment can be added to sandwiches and salads or served alongside meats to enhance flavors and add a nutritional boost.

Kimchi, known as a spicy fermented cabbage dish from Korea, is another versatile recipe ideal for preppers. Ingredients can include Napa cabbage, radish, scallions, garlic, ginger, and a mix of salt and Korean chili flakes for seasoning. The process is similar to making sauerkraut but includes a seasoning paste that gives kimchi its distinctive taste. Fermented for several days, kimchi can then be used as a side dish, added to soups or stews, or even used as a base for kimchi fried rice, providing a variety of meal options from a single ingredient.

A basic brine recipe can be the foundation for preserving a wide range of vegetables for those looking to pickle. A standard brine is made from water, vinegar, salt, and sugar, heated until the salt and sugar dissolve. This liquid can then be poured over prepared vegetables—such as cucumbers, carrots, and beans—in jars, sealing them for storage. These pickles can serve as crunchy snacks, flavorful additions to salads, or tangy accompaniments to main dishes.

Fermented beverages like kombucha and water kefir offer refreshing alternatives to store-bought drinks and are simple to prepare at home. Kombucha starts with sweetened tea that, when combined with a SCOBY (symbiotic culture of bacteria and yeast), ferments over the course of several days to a few weeks. Similarly, water kefir is made by fermenting sugar water with water kefir grains. Both beverages can be flavored with fruits, herbs, or spices and offer a probiotic-rich drink option.

Incorporating these pickled and fermented items into meals can be both creative and nutritious. A breakfast idea might include a slice of sourdough (another fermented favorite) topped with avocado and sauerkraut. For lunch, a salad dressed with pickled beets and carrots provides a flavorful and healthful option. Dinner could feature a kimchi stew or a side of kimchi with grilled meats. Even snacks can be elevated with homemade pickles or a glass of kombucha.

Storage and rotation are key for preppers using pickling and fermentation as part of their food strategy. Properly stored, pickled, and fermented foods can last for months, even years, but they should be rotated to ensure they are consumed at their peak quality. Labeling jars with the production date can help keep track of what should be used first.

Safety is paramount in the pickling and fermentation process. Preppers should ensure they follow trusted recipes and maintain a clean working environment to hinder contamination. Understanding the signs of spoilage, such as off smells, colors, or textures, is crucial to avoid consuming compromised foods.

In conclusion, for preppers, pickling and fermentation extend the shelf life of perishable foods and enhance their diet's flavor and nutritional profile. Through a combination of basic recipes and creative meal planning, these preservation methods can significantly contribute to a resilient and varied food supply. Whether enjoying the tangy crunch of a freshly opened jar of pickles or the comforting warmth of a kimchi stew, including pickled and fermented foods in a prepper's pantry offers practical and gourmet options for emergency food supplies.

# CHAPTER X

# Beyond the Basics: Creative Applications

### Pickling and fermenting unconventional ingredients

The art of pickling and fermenting, traditions steeped in history, have predominantly focused on familiar ingredients like cucumbers, cabbages, and carrots. Yet, the versatility and adaptability of these methods beckon culinary adventurers to explore beyond the conventional, venturing into the realm of fermenting and pickling unconventional ingredients. This section delves into the innovative practices of pickling and fermenting such ingredients, exploring not only the expansion of flavors and textures these methods offer but also their nutritional benefits and contributions to culinary diversity.

The process of pickling, which involves submerging ingredients in vinegar or brine, and fermentation, where natural bacteria ferment sugars into acids, alcohol, or gases, are both ancient food preservation techniques. These methods extend the shelf life of foods, enhance their nutritional profile, and introduce unique flavors and textures. When applied to unconventional ingredients, these techniques open up a world of culinary possibilities, encouraging creativity and experimentation in the kitchen.

One intriguing area of exploration is the pickling and fermenting of fruits. While certain fruits like apples and pears have traditionally been pickled, culinary adventurers are now turning their attention to tropical fruits such as pineapples, mangos, and papayas. Fermented pineapple, known in some cultures as tepache, transforms the fruit into a bubbly, tangy beverage, rich in probiotics. Similarly, pickled mangos can add a sweet, sour, and spicy kick to salads and salsas, enhancing dishes with complex flavors.

Another unconventional ingredient suitable for fermentation is dairy. Beyond the familiar realms of yogurt and kefir, fermenting various types of milk can yield a range of flavorful and nutritious products. Camel milk, for instance, can be fermented into a unique version of kefir, offering a different nutritional profile and flavor distinct from its cow milk counterpart. Such practices broaden the spectrum of fermented dairy products and introduce these traditional techniques to new audiences.

Vegetables that are typically overlooked in the context of pickling and fermenting also offer untapped potential. For example, fermenting sweet potatoes can yield a range of products from tangy, fizzy beverages to sour, crunchy condiments. While less common due to the fruit's propensity to brown, pickled avocado can be a delightful addition to tacos and sandwiches when prepared correctly, offering a unique texture and flavor profile.

The fermentation of grains and legumes is another area ripe for exploration. While fermented soy products like tempeh and miso are well-known, other grains and legumes can also benefit from fermentation. Fermented rice, often used in traditional Asian cuisines to make dishes like sake or rice wine, can also be used as a base for porridges or added to soups to enhance flavor and nutritional value. Similarly, fermenting lentils can increase their digestibility and nutritional availability, transforming them into savory pancakes or bread alternatives.

Nuts and seeds, while not traditionally considered for pickling or fermentation, can also be transformed through these processes. Fermented nut cheeses, made by culturing cashews, almonds, or macadamias with probiotic bacteria, offer a dairy-free alternative to traditional cheese, packed with flavors that can range from sharp and tangy to mild and creamy. Pickled pumpkin seeds, spiced with herbs and vinegar, can become a flavorful, crunchy snack or garnish.

Incorporating these unconventional ingredients into pickling and fermentation practices expands the culinary repertoire and contributes to a more sustainable food system. Utilizing a broader range of ingredients, especially those that might otherwise go to waste, can help minimize food waste and promote a more diverse diet. Additionally, fermenting and pickling can enhance the nutritional value of foods, making vitamins and minerals more bioavailable and introducing beneficial probiotics into the diet.

However, when venturing into the pickling and fermenting of unconventional ingredients, it is crucial to be mindful of food safety. Proper sanitation, understanding the signs of successful fermentation, and being aware of potential allergens are essential to ensure that these culinary experiments remain safe and enjoyable to consume.

In conclusion, exploring pickling and fermenting unconventional ingredients represents a fertile ground for culinary innovation. These methods allow for the extension of food preservation and encourage a deeper engagement with the diversity of flavors, textures, and nutritional benefits our food system offers. By pushing the boundaries of traditional pickling and fermenting, culinary enthusiasts can discover new tastes, contribute to food sustainability, and enrich our collective dining experience. As we continue to explore these ancient techniques with modern creativity, we reaffirm the timeless value of fermentation and pickling in enriching our diets and connecting us to the vast tapestry of global food culture.

## Fusion recipes incorporating pickled and fermented foods

In the culinary world, fusion cuisine represents the creative blending of culinary traditions from different cultures, producing innovative and sometimes surprising flavor combinations. Within this innovative realm, pickled and fermented foods, with their complex flavors and health benefits, have become key ingredients in developing fusion recipes. These ingredients add depth and tanginess to dishes and introduce probiotics into our diets, promoting gut health. This section explores the world of fusion recipes incorporating pickled and fermented foods, showcasing the versatility and global appeal of these ancient preservation methods.

One standout example of fusion incorporating fermented foods is the Kimchi Taco, a delightful blend of Korean and Mexican culinary traditions. Here, the spicy, tangy crunch of kimchi is used as a topping for tacos, a staple of Mexican cuisine. The kimchi complements the rich flavors of grilled meat, whether it's traditional beef, chicken, or a vegetarian substitute like grilled tofu, adding a layer of complexity and a burst of flavor that enhances the overall dish. The tacos are further adorned with a drizzle of sesame-soy dressing and a sprinkle of cilantro, marrying the distinct flavors of Korea and Mexico in each bite.

Another innovative fusion recipe is Sauerkraut Pizza, which might sound unconventional at first but results in a surprisingly delicious combination. In this dish, sauerkraut is used as a topping alongside traditional pizza ingredients such as tomato sauce, mozzarella cheese, and pepperoni or mushrooms. Before adding it to the pizza, the sauerkraut can be lightly sautéed with garlic and a touch of caraway seeds to mellow its sharpness. The richness of the cheese and meat is balanced by the tanginess of the sauerkraut, offering a unique flavor profile and a crispy texture that makes the pizza stand out.

Fermented foods can also be incorporated into desserts, as seen in the Kefir Cheesecake. This recipe uses kefir, a fermented milk drink similar to yogurt but with a more diverse set of probiotics, as a substitute for some of the cream cheese typically used in cheesecake. The kefir reduces the dessert's fat content and adds a slight tanginess that complements the sweetness of the cake. Topped with a compote made from pickled cherries, this cheesecake becomes a testament to the versatility of fermented and pickled ingredients in enhancing even sweet dishes.

Another area ripe for exploration is the fusion of pickled vegetables into Asian-inspired noodle dishes. For instance, Pickled Vegetable Ramen introduces a bright, acidic contrast to traditional ramen's rich, savory broth. Pickled radishes, carrots, and cucumbers, along with a soft-boiled egg, slices of roasted pork or tofu, and a generous helping of noodles, create a harmonious blend of flavors and textures. The pickled vegetables add a pop of color and a refreshing crunch and zesty flavor that elevates the dish.

Lastly, the fusion concept extends to breakfast dishes with the introduction of Fermented Berry Pancakes. In this recipe, a traditional pancake batter is enriched with a spoonful of fermented berry puree, adding a nuanced depth of flavor and a minimal tanginess that balances the sweetness of the maple syrup topping. The fermented berries can be made by combining fresh berries with a bit of sugar and allowing them to ferment for a few days until bubbly. This technique not only preserves the seasonal bounty of berries but also enhances their nutritional value by introducing beneficial bacteria.

Incorporating pickled and fermented foods into fusion recipes offers endless possibilities for creativity and flavor experimentation. These ingredients, rooted in the preservation techniques of various cultures, bring their distinct tastes and textures and their health benefits to the fusion cuisine table. As chefs and home cooks continue to explore and innovate, integrating pickled and fermented foods into fusion dishes represents a celebration of global culinary traditions, a testament to the universal language of food that connects us all.

In conclusion, the world of fusion cuisine provides a vibrant canvas for the incorporation of pickled and fermented foods, allowing for creating dishes that are not only flavorful and innovative but also nutritious. By embracing these ingredients, fusion recipes celebrate the diversity of global culinary traditions, offering diners a unique gastronomic experience that transcends geographical and cultural boundaries. As we continue to explore and experiment with these combinations, we deepen our appreciation for the rich tapestry of flavors that pickling and fermentation bring to our tables.

## Crafting homemade condiments and sauces

Creating homemade condiments and sauces using pickled and fermented foods is an exploration of taste, tradition, and creativity. Deeply rooted in history, these culinary practices offer a palette of flavors that enhance the dining experience, transforming ordinary meals into extraordinary ones. This section delves into the craft of making homemade condiments and sauces from pickled and fermented ingredients, highlighting their unique characteristics, versatility, and the endless possibilities they present to both home cooks and culinary professionals.

Pickled and fermented foods possess complex flavors developed through the preservation process. Natural fermentation or the addition of vinegar creates a distinct tangy, sour, or umami taste. These flavors become the foundation of homemade condiments and sauces, providing a depth that cannot be achieved with fresh ingredients alone. The process of pickling and fermenting not only extends the shelf life of produce but also enriches it with probiotics and nutrients, adding a healthful boost to any dish they accompany.

One classic example of a homemade condiment crafted from fermented foods is kimchi mayo. This fusion of creamy mayonnaise with spicy, tangy kimchi creates a versatile condiment that can elevate sandwiches, burgers, and even sushi rolls. The process involves blending high-quality mayonnaise with finely chopped kimchi and a bit of the kimchi brine to achieve the desired consistency and flavor intensity. The result is a condiment that marries the richness of mayonnaise with the complex flavors of kimchi, offering a creamy texture with a kick of heat and a note of acidity.

Similarly, sauerkraut mustard brings a new twist to a traditional condiment. By incorporating finely minced sauerkraut and a splash of the fermenting liquid into homemade or store-bought mustard, the mustard is transformed into a condiment with added layers of flavor and texture. The tartness of the sauerkraut complements the spicy, pungent taste of the mustard, creating a condiment that pairs well with sausages, deli meats, and pretzels, adding a probiotic punch to classic dishes.

Another innovative condiment is pickled beet relish, a sweet and tangy addition to salads, sandwiches, and grilled meats. The relish is made by dicing pickled beets and combining them with ingredients such as red onions, capers, and a hint of orange zest, all bound together with a reduction of the pickling liquid. This relish adds a vibrant color to dishes and introduces a subtle sweetness and acidity that can balance richer flavors.

Fermented hot sauce represents yet another realm of possibilities, showcasing the versatility of fermented chilies. Fermenting fresh chilies with garlic, salt, and sometimes additional spices or vegetables creates a base that can be blended and strained to produce a hot sauce with unparalleled depth. The fermentation process mellows the chilies' heat while enhancing their natural flavors, resulting in a hot sauce that is both fiery and complex, perfect for an additional zesty kick to any meal.

Lastly, yogurt tzatziki sauce exemplifies how fermented dairy products can be used to craft refreshing, tangy condiments. Combining thick, strained yogurt with grated cucumber, garlic, lemon juice, as well as fresh herbs like dill or mint creates a cooling and flavorful sauce. The lactic acid fermentation of the yogurt not only contributes probiotics but also a creamy texture and a subtle tang that complements the crispness of the cucumber and the brightness of the lemon and herbs.

Crafting homemade condiments and sauces with pickled and fermented foods invites experimentation and personalization, allowing individuals to adjust flavors to their liking and explore new combinations. This creative process enhances the culinary experience and encourages a deeper appreciation for the traditions of pickling and fermentation, connecting us to the cultural heritage of these practices.

Moreover, creating condiments and sauces from pickled and fermented ingredients promotes sustainability by utilizing preserved produce and minimizing food waste. It allows cooks to capture and enjoy the seasonal bounty throughout the year, adding variety and excitement to the daily diet.

In conclusion, homemade condiments and sauces crafted from pickled and fermented foods represent a fusion of tradition, flavor, and nutrition. They offer a way to elevate ordinary meals, introduce probiotic benefits, and explore the rich tapestry of tastes that fermentation and pickling can bring to the table. Whether it's the spicy kick of kimchi mayo, the complex depth of fermented hot sauce, or the refreshing tang of yogurt tzatziki, these condiments and sauces embody the creativity and versatility of culinary arts, inviting us to experiment, taste, and enjoy.

## Innovations in pickling and fermentation technology

In recent years, the revival of interest in traditional food preservation methods such as pickling and fermentation has coincided with technological advancements, leading to significant innovations in this ancient culinary practice. These advancements have streamlined the process and expanded the possibilities for flavor development, safety, and nutritional enhancement. This section explores the cutting-edge innovations in pickling and fermentation technology, examining how they have transformed these time-honored methods into a modern science that continues to evolve and intrigue.

At the forefront of these innovations is the development of precision fermentation technology. This process involves genetically engineering microorganisms such as yeast, bacteria, or fungi to produce specific compounds through fermentation. Precision fermentation can be used to create ingredients that are identical to those found in animal products, such as proteins, fats, and flavors, without the need for traditional animal farming. This technology is particularly promising for producing sustainable, plant-based alternatives to dairy and meat products, offering a potential solution for some of the environmental and ethical concerns that are connected with animal agriculture.

Another significant advancement is the use of controlled fermentation environments. Modern fermentation chambers and bioreactors allow for precise control over temperature, humidity, and oxygen levels, conditions that can significantly affect the fermentation process's outcome. By fine-tuning these variables, producers can achieve consistent results, enhance specific flavors, and even speed up fermentation. These controlled environments also reduce the risk of contamination by unwanted microorganisms, ensuring the safety and quality of the final product.

Integrating digital technology into pickling and fermentation processes represents a further innovation. Smart fermentation devices equipped with sensors and connected to smartphone apps are now available for home use, enabling enthusiasts and artisans to monitor and control the fermentation process remotely. These devices can track temperature, pH levels, and other critical parameters, providing real-time data that can be used to adjust conditions as needed. This level of control and visibility makes the fermentation process more accessible to novices and allows experienced practitioners to experiment with greater precision.

Advancements in packaging technology have also been crucial in the evolution of pickling and fermentation. Vacuum-sealing techniques and the development of oxygen-absorbing and antimicrobial packaging materials can extend the shelf life of pickled and fermented products while preserving their taste and nutritional value. Additionally, using biodegradable and compostable packaging materials aligns with the sustainable ethos of many producers and consumers of pickled and fermented foods, addressing concerns about the environmental influence of food packaging waste.

Exploring novel ingredients and fermentation substrates has expanded the boundaries of traditional pickling and fermentation. Researchers and culinary innovators are experimenting with various fruits, vegetables, grains, and even dairy alternatives, uncovering new flavors and health benefits. The fermentation of plant-based milks, for example, has led to the development of vegan yogurts and cheeses that closely mimic the taste and texture of their dairy counterparts. Similarly, the pickling of uncommon fruits and vegetables has introduced consumers to diverse flavors and culinary possibilities.

Furthermore, applying fermentation technology to waste reduction is an exciting area of innovation. The fermentation of agricultural by-products and food waste can produce valuable products such as biofuels, animal feed, and organic acids, turning what would otherwise be discarded into useful resources. This contributes to a more sustainable food system and adds economic value for farmers and producers.

In conclusion, pickling and fermentation technology innovations are revolutionizing these ancient practices, making them more accessible, consistent, and versatile. These technological advancements offer promising solutions to contemporary challenges, including the need for sustainable food production, waste reduction, and the development of plant-based alternatives to animal products. As we continue to explore the intersection of tradition and technology, the future of pickling and fermentation holds exciting possibilities for flavor, health, and environmental sustainability. By embracing these innovations, we can carry forward the rich heritage of pickling and fermentation into a new era of culinary exploration and discovery.

# CONCLUSION

"Pickling and Fermentation for Preppers: Nourishment for Challenging Times" has been a journey of discovery and empowerment, guiding readers through the ancient art of pickling and fermentation to prepare for uncertain times. As we conclude this book, it's evident that the knowledge and skills shared within its pages hold immense value, not only in times of crisis but also in everyday life.

Throughout our exploration, we've uncovered the intricate science behind pickling and fermentation, gaining a deeper appreciation for the microbial processes that transform simple ingredients into complex flavors and probiotic-rich foods. From understanding the role of beneficial bacteria to mastering the factors that influence successful fermentation, readers have been equipped with the foundational knowledge needed to embark on their own fermentation journey.

The practical aspects of pickling and fermentation have also been thoroughly explored, from essential equipment and ingredients to various recipes for pickled vegetables, fruits, and beverages. Whether preserving the bounty of the harvest or simply adding a burst of flavor to everyday meals, these techniques offer endless possibilities for creativity and sustenance.

Moreover, the health benefits of pickled and fermented foods cannot be overstated. From supporting gut health to bolstering the immune system, these age-old preservation methods offer a wealth of advantages that are especially valuable in times of need. By incorporating these foods into their diet, readers can nourish their bodies and cultivate resilience and vitality.

As we look toward the future, it's clear that the principles of pickling and fermentation will continue to be vital in our quest for self-sufficiency and sustainability. Emerging trends and technologies promise to revolutionize the way we prepare and store food further, ensuring that future generations inherit a legacy of resilience and resourcefulness.

In closing, "Pickling and Fermentation for Preppers" is more than just a cookbook—it's a roadmap to self-reliance, empowerment, and nourishment in the face of adversity. Whether you're a seasoned fermenter or a novice prepper, I hope this book has inspired you to embrace the timeless art of pickling and fermentation, empowering you to take control of your food supply and thrive in even the most challenging times.

*Thank you for buying and reading/ listening to our book. If you found this book useful/ helpful please take a few minutes and leave a review on the platform where you purchased our book. Your feedback matters greatly to us.*